English Learner Support Guide

Mc
Graw
Hill
Education

Bothell, WA • Chicago, IL • Columbus, OH • New York, NY

Author
Sharon Griffin

Professor Emerita of Education and Psychology

Clark University
Worcester, Massachusetts

www.mheonline.com

Send all inquiries to:
McGraw-Hill Education
8787 Orion Place
Columbus, OH 43240

ISBN: 978-0-02-129986-7
MHID: 0-02-129986-2

Printed in the United States of America.

1 2 3 4 5 6 7 8 9 RHR 17 16 15 14 13

Contents

Contents

Introduction

This **English Learner Support Guide** provides both teacher and student support for **Number Worlds** to ensure access and ultimate success for all students in mathematics. The goal for English learners is to understand and use English in social as well as academic contexts so that they can fully participate in mathematics learning.

This guide enables teachers to make the Number Worlds lessons as comprehensible as possible for students learning English, while also providing opportunities for oral and written responses.

Students must develop such receptive English language skills as listening and reading and use them in turn to acquire vocabulary. Students must also develop their speaking skills to communicate what they have learned.

The following elements of English may present a particular challenge to students acquiring English:

▶ **English phonemes that may not exist in the student's home language**

▶ **Vocabulary**

▶ **Idioms and expressions**

▶ **Comprehension of written text**

The vocabulary instruction in the **English Learner Support Guide** will help students with general academic language development for receptive comprehension.

Academic Vocabulary words are words students need to know and be able to use for the upcoming week of **Number Worlds**. These words are essential for comprehending the lesson that students will encounter in future lessons. These words need to be taught explicitly, with multiple opportunities for students to practice using the words in context. In order for students to understand the mathematics lesson, they will need guidance in content-specific vocabulary. In addition, these meaningful vocabulary words will help the student's growth in English. The emphasis in the **Number Worlds English Learner Support Guide** is on developing the words needed for completing mathematics lessons with comprehension.

Resources

The *Number Worlds* program includes a variety of program materials designed to provide English learner support.

English Learner Support Guide

Teacher Edition

Student Edition

Vocabulary Cards

Each level of the *English Learner Support Guide* includes the following:

• Individual Oral Assessments

• Vocabulary Instruction

• Progress Monitoring

• Practice Pages for three levels of language acquisition (Levels D–J)

• Summative Assessments

• Blackline Masters

• Vocabulary Flashcards for additional practice

Stages of English Language Proficiency

An effective learning environment is an important goal of all teachers. In a supportive environment, all English learners have the opportunity to participate and to learn. The materials in this guide are designed to support students while they are acquiring English.

This guide provides direction in supporting students in three stages of English proficiency: Beginning, Intermediate, and Advanced. These three stages can be described in general terms as follows.

■ **Beginning** Students identified at this level of English-language proficiency know little English and will probably have difficulty comprehending English. During this stage, students demonstrate dramatic growth as they progress from having no receptive or productive English to possessing a basic command of English. They are learning to comprehend and produce one- or two-word responses to questions and are moving to phrases and simple sentences using concrete and immediate topics and to interact in a limited fashion with text that has been taught. They progress to responding with increasing ease to more varied communication tasks using learned material, comprehending a sequence of information on familiar topics, producing basic statements, asking questions on familiar subjects, and interacting with a variety of print. Many basic errors are found in their use of English syntax and grammar.

■ **Intermediate** Students who have reached the intermediate level of English-language proficiency have good comprehension of overall meaning and are beginning to demonstrate increased comprehension of specific details and concepts. They are learning to respond in expanded sentences, interact more independently with a variety of texts, and use newly acquired English vocabulary to communicate ideas orally and in writing. They demonstrate fewer errors in English grammar and syntax than at the Beginning level.

■ **Early Advanced/Advanced** Students who are identified at this level of English-language proficiency demonstrate consistent comprehension of meaning, including implied and nuanced meaning, and are learning the use of idiomatic and figurative language. They are increasingly able to respond using detail in compound and complex sentences and to sustain conversations in English. They are able to use standard grammar with few errors and show understanding of conventions of formal and informal usage.

Academic Language Acquisition

English learners often sound fluent in conversation, which leads teachers to believe that they are completely proficient in English in all settings. Social conversation is often concrete, discussing events and subjects familiar to English learners. However, academic language, the language that allows students to discuss abstract concepts, employs sophisticated grammatical constructions. Proficiency in communication using academic language is important for the success of all students, including English learners.

Mathematics, like all content-area subjects, has a special language that helps students learn key concepts and discuss them with others. This academic language includes unique terminology to mathematics, as well as the signs and symbols that are used to convey common understandings.

The academic language of mathematics is not usually encountered in social settings. It must be explicitly taught and practiced. One unique challenge to teachers of English learners is that students learning English may arrive at school for the first time in any grade level. This means that, at every grade level, teachers need to be prepared with a tool kit of strategies to accelerate language learning and to backfill skills that may have been taught and mastered by other students in previous grade levels.

Another challenge is that some students arrive with high levels of mathematics preparation. These students may know the mathematics but now must learn the subject matter discourse in order to ask questions, read and comprehend the textbook, and otherwise participate in class.

Ways to Teach Vocabulary

The Creating Context suggestion in each lesson in the **Number Worlds Teacher Edition** provides teachers with valuable tools to meet the needs of English learners. Often English learners understand the mathematics in a lesson but cannot communicate that understanding in English. The following are some strategies for teaching vocabulary that can be incorporated in any subject.

■ **Real Objects and Realia** Because of the immediate result visuals have on learning language, the best approach to explain a word such as *car* is simply to show a real car. As an alternate to the real object, you can show realia. Realia are toy versions of real things, such as plastic eggs to substitute for real eggs, or in this case, a toy car to signify a real car. A large, clear picture of an automobile can also work if it is absolutely recognizable.

If, however, the student has had no experience with the item in the picture, more explanation might be needed. For example, if the word you are explaining is a zoo animal such as an ocelot and the students are not familiar with this animal, one picture might be insufficient. They might confuse this animal with a cat or any one of the feline species, such as a tiger. Seeing several clear pictures, then, of each individual type of common feline and comparing their similarities and differences might help clarify meaning in this particular instance. When students make a connection between their prior knowledge of the word *cat* with the new word *ocelot,* it validates their newly acquired knowledge, and thus they process learning more quickly.

■ **Pictures** In many cases bringing real objects into the classroom is not practical. Supplement illustrations with visuals such as those found in the **SRA Photo Library,** magazine pictures, and picture dictionaries. Videos, especially those that demonstrate an entire setting such as a farm or a zoo, or videos where different animals are highlighted in their natural habitat, for instance, might be helpful. You might also wish to turn off the sound to avoid a flood of language that students might not be able to understand. Since you want them to concentrate in the visual/word-meaning correlation, leaving the sound on will take away meaning from the visual/interpretative connection. Students generally will be able to establish links more easily without sound, especially at the beginning of English language acquisition.

Using drawings that are sketchy, indistinct, or incomplete may be cryptic to students. When possible, select large, clear, and simple drawings that show entire objects, or use clear photos.

Ways to Teach Vocabulary

■ **Graphic Organizers** Various kinds of graphic organizers and semantic maps can be very helpful, particularly with students who have attained Intermediate English-language proficiency and for moving students to higher levels of proficiency.

■ **Pantomime** Language is learned through modeling within a communicative context. Pantomiming is one example of such a framework of communication. Some words, such as *run* or *jump,* are appropriate for pantomiming. You could use photos or picture cards for verbs like these, but demonstrating running and jumping is also necessary to solidify meaning. If students understand what you are trying to pantomime or if they recognize what it is you are striving to signify through your gestures or your facial expressions, they will more easily engage in the task of learning.

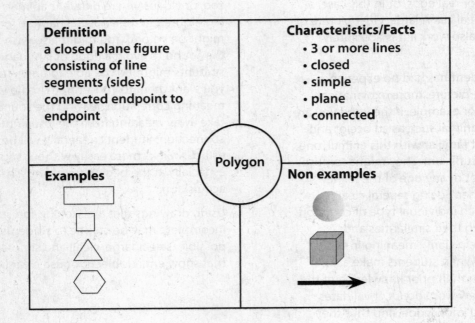

Sheltered Instruction

When English learners work in the core-content areas, they face the challenge of both the new concept load and the new language load. Sheltered instruction designs lesson delivery to take advantage of student language strengths and techniques that do not rely solely on telling, but on doing so that students see, hear, and experience the new concept along with the new language. **Number Worlds** employs the following sheltered instruction strategies throughout the program.

■ **Accessing Prior Knowledge** Sheltered Instruction begins with accessing students' prior knowledge to determine what they already know and how much background building may be necessary. The lessons in **Number Worlds** begin with Warm Up exercises and follow up with discussions and activities that provoke students to think about and share what they already know, allowing the teacher to diagnose where gaps may exist.

■ **Modeling** Demonstrating new math skills and concepts is a crucial developmental focus of **Number Worlds**, and it is one of the key strategies of Sheltered Instruction. When new concepts and skills are introduced, prior knowledge provides scaffolding as students see their teacher use manipulatives, hear appropriate terminology, and practice setting.

■ **Explicit Instruction** The step-by-step instruction in **Number Worlds** is carefully structured to lead students to understanding the concepts. Built into the teaching are explanations of new terms and skills. It provides for plenty of repetition and revisiting of these concepts using the vocabulary so that students have ample modeling before they are expected to use these terms themselves.

■ **Checking for Understanding** Sheltered Instruction involves interaction with the teacher, including checking for understanding along the way. When students are in the early stages of English language acquisition, they may not always be the first to answer a question or be able to fully express their ideas in speech or in writing. **Number Worlds** builds in many modes of checking comprehension throughout the lesson so that all students can show what they know. For example, a lesson may suggest that students can use a signal, such as thumbs-up or thumbs-down, to indicate their understanding. In this way, English learners access the core concepts but do not have to rely only on English expression to show what they know or where they need additional help.

Sheltered Instruction

■ **Math Games** English learners need to practice their new language in authentic settings rather than with drills that do not capture the language tasks students need to perform in school. Language learning can be a high-anxiety experience because the nature of the learning means that the speaker will make many mistakes. This can be profoundly difficult in front of one's peers or in answering individual questions in front of the whole class. Working in pairs and groups allows English learners a way to understand in a low-anxiety environment and allows more learning to take place.

A big part of the practice built into *Number Worlds* is in the form of learning games. In addition to being fun and motivating, math games provide English learners additional opportunity to practice. Vocabulary is woven into appropriate language repetition and revisiting skills and strategies. In addition, these specially designed games model strategic thinking so that even students at beginning levels of proficiency learn strategy and can demonstrate strategic, higher-level thinking.

■ **Preview/Review Technique** A quick review of the key concepts in the primary language allows students to converse with peers who speak the same primary language to be sure concepts are clear. If the teacher or an aide speaks the primary language of the students, this provides the additional benefit of answering questions and elaborating on the key concepts for students.

One of the true benefits of Preview/Review is that students focus on the lesson in English rather than waiting to alternate with the primary language translation. The teacher can use the English words in every lesson as an example of what the primary language preview should include.

Strategies for Teaching English Learners

Strategy 1

Build a classroom environment to ensure success.

- Make sure students see the names of objects. Label everything.
- Speak clearly without distorting language.
- Learn words in your students' home language. Share them with the class, introducing their culture to all.
- Assign helper duties to your English learners to ensure they feel part of the class.
- Make sure to praise students often, creating opportunities for success.

Strategy 2

Build on prior knowledge.

- Give English learners a list of vocabulary words before they use the words in lessons. Provide a dictionary to help students learn the words.
- Use Venn diagrams, KWL charts (What I Know, What I Want to Know, What I Learned), and other graphic organizers to help students relate words.
- Allow discussion before assigning a lesson to help students build linguistic skills.

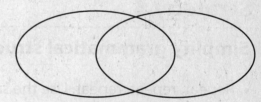

Use Venn diagrams to sort objects and characteristics.

Strategy 3

Incorporate culture at every opportunity.

- Provide cultural background for students. Often, text assumes cultural familiarity.
- Include relevant examples whenever possible. Comparing situations to those in another culture may help.

Strategies for Teaching English Learners

Strategy 4

Give visual clues.

- Take vocabulary words out of the written context and provide visuals clue to help build students' understanding. Use pictures, charts, and graphic organizers as clues.
- Use graphic organizers, such as Venn diagrams, to build visual associations.

	Number of sides	Number of angles or corners	Number of right angles	Notes
Square				
Triangle				
Circle				
Rectangle				

Strategy 5

Simplify grammatical structures.

- Repeat, repeat, repeat. Use the same sentence structure when giving directions.
- Summarize lessons in simpler form.
- Introduce one concept per sentence.
- Simplify sentences, but include essential academic vocabulary.
- Use the subject-verb-object sentence structure.
- Use the active, not passive, voice.

Strategy 6

Teach vocabulary while teaching the lesson.

- Model pronunciation.
- Create an environment where students feel safe to make a mistake.
- Explain idiomatic language and word meanings.

Strategy 7

Teach for depth of understanding.

- Develop a clear, in depth understanding of content.
- Focus on learning a few concepts well to give English learners a greater opportunity for success.

Strategy 8

Teach study skills.

- Guide learners to use text resources: tables of contents, glossaries, indexes, captions, and stylized text.
- Help students who cannot read the text use words and pictures to categorize the text.
- Make sure that learners realize that textbooks are written to provide information.
- Organize students into pairs as learning companions.
- Provide students with categorical word lists.
- Create topic outlines.

Vocabulary
- **digit** Any of the Arabic numerals 0 through 9
- **expanded form** A way to write a number that shows the value of each digit; for example, $300 + 50 + 6$
- **standard form** A number written using the digits 0–9; for example, *178*
- **word form** A number written using words; for example, *sixty three*

Strategy 9

Use manipulatives.

- Involve English learners with hands-on activities.
- Encourage students to construct, graph, measure, and experiment.
- Encourage discussion while students use hands-on materials.

Strategies for Teaching English Learners

Strategy 10

Inform and adapt instruction.

- Ask questions based on students English proficiency level.

- Monitor progress daily.

- Allow students to show their knowledge in different ways, especially by using visual aids, graphs, and drawings.

Progress Monitoring	
If... students need additional practice using coordinate grids,	**Then...** use coordinates to find a place on a city map. Using a map of your school or municipality will give English learners additional practice recognizing and pronouncing street names where they live.

Strategy 11

Use cooperative learning.

- Use the buddy system to help learners work efficiently. English learners working cooperatively make substantial learning gains.

- Include English learners in group work. Even if their comprehension is low, they will learn academic vocabulary and skills.

Lesson Routines

1 WARM UP

Begin each day with the lesson Warm Up with the entire group. These short activities are used in a whole-group format at the beginning of each day's lesson.

Warm Up is an essential component because it helps students to preview vocabulary they will use in the **Engage** activities. It also gives students daily opportunities to sharpen their vocabulary understanding.

2 ENGAGE

Engage includes intensive vocabulary instruction. Suggestions are included to introduce vocabulary. Guided discussion, game demonstrations, and strategy-building activities develop student understanding. Before beginning the activity, familiarize yourself with the lesson vocabulary, while making sure materials are available for student use. Introduce the activity, and use discussion to help students explore the lesson vocabulary.

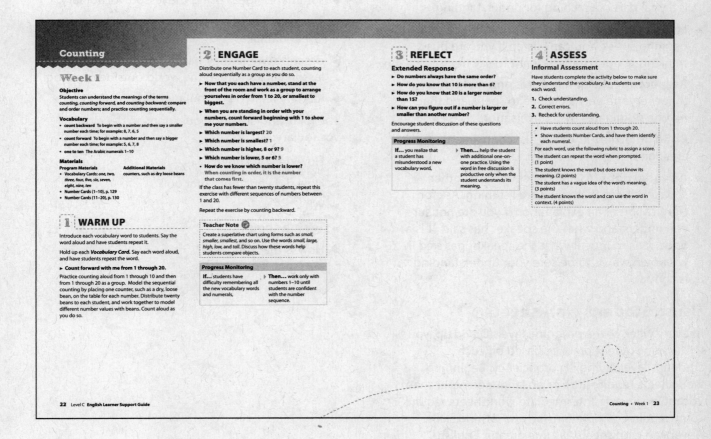

Counting

Week 1

Objective
Students can understand the meanings of the terms *counting, counting forward,* and *counting backward*; compare and order numbers; and practice counting sequentially.

Vocabulary
- **count backward** To begin with a number and then say a smaller number each time; for example: 8, 7, 6, 5
- **count forward** To begin with a number and then say a bigger number each time; for example: 5, 6, 7, 8
- **one to ten** The Arabic numerals 1–10

Materials

Program Materials	Additional Materials
• Vocabulary Cards: *one, two, three, four, five, six, seven, eight, nine, ten*	counters, such as dry loose beans
• Number Cards (1–10), p. 129	
• Number Cards (11–20), p. 130	

1 WARM UP

Introduce each vocabulary word to students. Say the word aloud and have students repeat it.

Hold up each *Vocabulary Card.* Say each word aloud, and have students repeat the word.

► **Count forward with me from 1 through 20.**

Practice counting aloud from 1 through 10 and then from 1 through 20 as a group. Model the sequential counting by placing one counter, such as a dry, loose bean, on the table for each number. Distribute twenty beans to each student, and work together to model different number values with beans. Count aloud as you do so.

2 ENGAGE

Distribute one Number Card to each student, counting aloud sequentially as a group as you do so.

► Now that you each have a number, stand at the front of the room and work as a group to arrange yourselves in order from 1 to 20, or smallest to biggest.

► When you are standing in order with your numbers, count forward beginning with 1 to show me your numbers.

► Which number is largest? 20

► Which number is smallest? 1

► Which number is higher, 8 or 9? 9

► Which number is lower, 5 or 6? 5

• How do we know which number is lower? When counting in order, it is the number that comes first.

If the class has fewer than twenty students, repeat this exercise with different sequences of numbers between 1 and 20.

Repeat the exercise by counting backward.

Teacher Note

Create a superlative chart using forms such as *small, smaller, smallest,* and so on. Use the words *small, large, high, low,* and *tall.* Discuss how these words help students compare objects.

Progress Monitoring

If... students have difficulty remembering all the new vocabulary words and numerals,	► Then... work only with numbers 1–10 until students are confident with the number sequence.

3 REFLECT

Extended Response

► Do numbers always have the same order?

► How do you know that 10 is more than 6?

► How do you know that 20 is a larger number than 15?

► How can you figure out if a number is larger or smaller than another number?

Encourage student discussion of these questions and answers.

Progress Monitoring

If... you realize that a student has misunderstood a new vocabulary word,	► Then... help the student with additional one-on-one practice. Using the word in free discussion is productive only when the student understands its meaning.

4 ASSESS

Informal Assessment

Have students complete the activity below to make sure they understand the vocabulary. As students use each word:

1. Check understanding.
2. Correct errors.
3. Recheck for understanding.

- Have students count aloud from 1 through 20.
- Show students Number Cards, and have them identify each numeral.

For each word, use the following rubric to assign a score.
The student can repeat the word when prompted. (1 point)
The student knows the word but does not know its meaning. (2 points)
The student has a vague idea of the word's meaning. (3 points)
The student knows the word and can use the word in context. (4 points)

Lesson Routines

Classroom discussion is encouraged in every lesson. Rules for classroom discussion established at the beginning of the year make each lesson more meaningful and promote listening and speaking skills. The following are some helpful ways to facilitate classroom discussion.

▶ **Pay attention to others.** Give your full attention to the person who is speaking. This includes looking at the speaker and nodding to show that you understand.

▶ **Wait for speakers to answer and complete their thoughts.** Sometimes teachers and other students get impatient and move on to ask someone else before someone has a chance to think and speak. Giving students time to answer is a vital part of teaching for understanding.

▶ **Listen.** Let yourself finish listening before you begin to speak. You can't listen if you are busy thinking about what you want to say next.

▶ **Respect speakers.** Take turns and make sure that everyone gets a chance to speak and that no one dominates the conversation.

▶ **Build on others' ideas.** Make connections, draw analogies, or expand on the idea.

▶ **Ask questions.** Asking questions of another speaker shows that you were listening. Ask for clarification or an explanation if you are not sure you understand what the speaker has said. It is a good idea to repeat what the speaker has said in your own words to be sure your understanding is correct.

Using Student Worksheets

In every week of the program, levels D and up, you will assign student practice based on each student's level of English proficiency. Beginning worksheets require little or no language other than vocabulary words. Intermediate worksheets require some reading comprehension and writing skills. Advanced worksheets assume greater English-language proficiency. At each level of English proficiency, the worksheet pages help students learn to use in context the vocabulary presented throughout the week.

3 REFLECT

Reflect is a vital part of the lesson that offers ways to help students summarize and reflect on their understanding of lesson vocabulary. When students talk about their thinking, they engage in mathematical generalizing and communicating. Allowing students to discuss what they did during an activity helps build mathematical reasoning and also develops social skills, such as taking turns, listening, and speaking. At the designated time, have students stop working and direct their attention to reflecting on the lesson.

Use the suggested questions in **Reflect** or have students

▶ summarize ideas using the lesson vocabulary.

▶ compare how the lesson vocabulary words are like and unlike.

▶ describe where they have seen or can apply the lesson vocabulary in the world outside of school.

▶ identify related vocabulary.

The following are good questions to develop vocabulary:

▶ How do you know?

▶ How did you figure that out?

▶ Why?

▶ Tell me about…

▶ How is that the same?

Lesson Routines

4 ASSESS

Assess helps you use informal and formal assessments to summarize and analyze evidence of student understanding and plan for differentiating instruction.

Progress Monitoring

Informal daily assessments evaluate students' vocabulary proficiency. **Warm Up** exercises and **Engage** activities can be used for day-to-day observation and assessment of how well each student is learning skills and grasping concepts. The **Engage** activities allow you to watch students using vocabulary under conditions more natural to them than most classroom activities. **Warm Up** exercises allow you to see individual responses, give immediate feedback, and involve the entire class.

Formal Assessment

Final Assessments provide formal assessments for each week of the *Number Worlds* program. These assessments provide information about the level of understanding of each of the vocabulary words taught in the *Number Worlds* program. They also help teachers continually monitor the student's level of English language proficiency in the context of content-area mathematics.

Weeks 1–5

Section at a Glance

In this section, students will learn the vocabulary associated with **Number Worlds,** Level C, Weeks 1–5. Students are expected to count to 100; compare and order numbers; understand *more* and *equal*; solve missing addend problems; understand the meaning of *enough, too many,* and *too few;* identify and sequence numerals to 20; and identify or compute set size. Before beginning the section, assess students' general knowledge of math vocabulary using the Individual Oral Assessment on page 21.

How Students Learn Vocabulary

By counting and identifying sets of tangible objects, English learners develop recognition of numbers and gradually understand that each number occupies a fixed position in a number sequence. Relating concepts of *enough, many, few, more,* and *equal* to daily life outside the classroom will help English learners develop understanding of these concepts in mathematics.

Academic Vocabulary Taught in Weeks 1–5

Week 1

count backward To begin with a number and then say a smaller number each time; for example: 8, 7, 6, 5

count forward To begin with a number and then say a bigger number each time; for example: 5, 6, 7, 8

one to ten The Arabic numerals 1–10

Week 2

compare To think about how things are alike and different

equal Having the same amount

least The smallest amount

less A smaller number or amount

more A larger number or amount

most The largest amount

numeral A symbol that represents a number

Week 3

compare To think about how things are alike and different

enough The correct amount

equal Having the same amount

few Made up of a small number

many Made up of a large number

numeral A symbol that represents a number

Week 4

higher number/lower number A higher number has more, a lower number has less

match To find two things that are the same

numeral A symbol that represents a number

symbol A letter, number, or picture that stands for something else

Week 5

after Behind; to come later

before In front of; to come first

far Not close to; a long way away

line A straight path that extends infinitely in opposite directions, thought of as having length, but no thickness

near Close to

Weeks 1–5 Individual Oral Assessment

Directions: Read each question to the student, and record his or her oral responses. Some questions have teacher directions. Teacher directions are indicated in italics. Allow students to use pencil and paper to work their responses.

1. I will **count forward** aloud. Say *yes* if I have counted correctly or *no* if I have missed any numbers. *Count aloud from 1 to 10. Say each number.* **yes**

2. I will **count forward** aloud. Say *yes* if I have counted correctly or *no* if I have missed any numbers. *Count aloud from 1 to 20. Say each number.* **yes**

3. I will **count forward** aloud. Say *yes* if I have counted correctly or *no* if I have missed any numbers. *Count aloud from 1 to 10. Skip the number 8.* **no**

4. I will **count backward.** Tell me if I have counted correctly or if I have missed any numbers. *Count aloud from 10 to 1. Say each number.* **Yes, you counted correctly.**

5. I will **count backward.** Tell me if I have counted correctly or if I have missed any numbers. *Count aloud from 10 to 1. Skip the number 5.* **No, you skipped 5.**

6. What number is this, **six** or **nine?** *Write the numeral 6 on the board.* **6**

7. Which set of counters has **more?** *Display two sets of counters, one with eight counters and one with four counters.* **the set with eight counters**

8. Which set of counters has the **least** amount? *Refer to the two groups of counters above.* **the set with four counters**

9. **Compare** these two sets of counters. Are they **equal?** *Display two sets of seven counters.* **yes**

10. How do you know that they are **equal?** **I counted them, and they have the same number.**

11. Which number is **higher?** *Write 5 and 9 on a piece of paper.* **9**

12. Do you put your socks on **before** or **after** you put on your shoes? **before**

13. Do I have **enough** paper clips to put one in each circle? *Draw ten circles on a piece of paper, and display six paper clips.* **no**

- **Beginning English Learners:** 0–3 of Questions 1–10 correct
- **Intermediate English Learners:** 4–7 of Questions 1–10 correct
- **Advanced English Learners:** 8–10 of Questions 1–10 correct
- If the student is able to answer Questions 11–13, then he or she can understand the mathematics taught in this unit but may still have difficulty with the academic vocabulary.

Use the Student Assessment Record, page 142, to record the assessment results.

Counting

Week 1

Objective

Students can understand the meanings of the terms *counting, counting forward,* and *counting backward;* compare and order numbers; and practice counting sequentially.

Vocabulary

- **count backward** To begin with a number and then say a smaller number each time; for example: 8, 7, 6, 5
- **count forward** To begin with a number and then say a bigger number each time; for example: 5, 6, 7, 8
- **one to ten** The Arabic numerals 1–10

Materials

Program Materials
- Vocabulary Cards: *one, two, three, four, five, six, seven, eight, nine, ten*
- Number Cards (1–10), p. 129
- Number Cards (11–20), p. 130

Additional Materials
counters, such as dry loose beans

1 WARM UP

Introduce each vocabulary word to students. Say the word aloud and have students repeat it.

Hold up each *Vocabulary Card.* Say each word aloud, and have students repeat the word.

▶ **Count forward with me from 1 through 20.**

Practice counting aloud from 1 through 10 and then from 1 through 20 as a group. Model the sequential counting by placing one counter, such as a dry, loose bean, on the table for each number. Distribute twenty beans to each student, and work together to model different number values with beans. Count aloud as you do so.

2 ENGAGE

Distribute one Number Card to each student, counting aloud sequentially as a group as you do so.

- ▶ **Now that you each have a number, stand at the front of the room and work as a group to arrange yourselves in order from 1 to 20, or smallest to biggest.**
- ▶ **When you are standing in order with your numbers, count forward beginning with 1 to show me your numbers.**
- ▶ **Which number is largest? 20**
- ▶ **Which number is smallest? 1**
- ▶ **Which number is higher, 8 or 9? 9**
- ▶ **Which number is lower, 5 or 6? 5**
- • **How do we know which number is lower?** When counting in order, it is the number that comes first.

If the class has fewer than twenty students, repeat this exercise with different sequences of numbers between 1 and 20.

Repeat the exercise by counting backward.

Teacher Note

Create a superlative chart using forms such as *small, smaller, smallest,* and so on. Use the words *small, large, high, low,* and *tall.* Discuss how these words help students compare objects.

Progress Monitoring

If... students have difficulty remembering all the new vocabulary words and numerals,	▶ Then... work only with numbers 1–10 until students are confident with the number sequence.

3 REFLECT

Extended Response

▶ Do numbers always have the same order?

▶ How do you know that 10 is more than 6?

▶ How do you know that 20 is a larger number than 15?

▶ How can you figure out if a number is larger or smaller than another number?

Encourage student discussion of these questions and answers.

Progress Monitoring

If... you realize that a student has misunderstood a new vocabulary word,	▶ Then... help the student with additional one-on-one practice. Using the word in free discussion is productive only when the student understands its meaning.

4 ASSESS

Informal Assessment

Have students complete the activity below to make sure they understand the vocabulary. As students use each word:

1. Check understanding.

2. Correct errors.

3. Recheck for understanding.

- Have students count aloud from 1 through 20.
- Show students Number Cards, and have them identify each numeral.

For each word, use the following rubric to assign a score.

The student can repeat the word when prompted. (1 point)

The student knows the word but does not know its meaning. (2 points)

The student has a vague idea of the word's meaning. (3 points)

The student knows the word and can use the word in context. (4 points)

Counting and Comparing

Week 2

Objective
Students can understand the meanings of the terms *more* and *equal* and can compare number values.

Vocabulary
- **compare** To think about how things are alike and different
- **equal** Having the same amount
- **least** The smallest amount
- **less** A smaller number or amount
- **more** A larger number or amount
- **most** The largest amount
- **numeral** A symbol that represents a number

Materials
Program Materials
Vocabulary Card: *equal*

Additional Materials
- counters, such as dry beans
- index cards
- paper plates

1 WARM UP

Introduce each vocabulary word to students. Say the word aloud and have students repeat the word.

After defining the vocabulary words, show students the *equal* **Vocabulary Card 13** while saying the word aloud.

Practice counting aloud as a group and individually from 1 to 20. Organize students into pairs, and give each pair of students a paper plate. As a class, count aloud as you place different numbers of dry beans onto each plate. Distribute the dry beans in values of 1 to 20. Distribute an index card to each pair of students.

Have each pair of students write the number of dry beans it has on an index card to label the plate. Have each pair use the sentence *We have _____ beans* to describe the value.

2 ENGAGE

Organize pairs into groups of four.

▶ **Use these sentences to compare your beans:**
- **We have _____ beans.**
- **The other pair has _____ beans.**
- **The other pair has _____ (more or less) beans than we have.**
- **We have _____ (more or less) beans than the other pair.**

▶ **Who has more beans?** Students should identify the pair with the greater number of beans.

▶ **Look at all the plates of beans. Who has the most beans? Who has the least number of beans?** Students should indicate the plate with the greatest number of beans and the plate with the least number of beans.

▶ **Does anyone have equal numbers of beans?** Students should indicate two plates with the same number of beans.

▶ **How many beans would your pair need to be equal to the other pair's plate of beans?** Answers will vary.

Ask students to combine two plates of beans by counting sequentially for the first plate and continuing the sequence by counting the second plate. Find the total number of beans in the class by sequentially counting each plate of beans.

Teacher Note

To determine whether students understand the concepts of most and least, ask them to point to the part of the classroom with the most desks, the least number of students, and so on. This will allow students at beginning levels of proficiency to participate before you launch the math instruction on comparing numbers.

Progress Monitoring

If... students are comfortable with values between 1 and 20,	▶ **Then...** use greater numbers within the group's comfort zone.

3 REFLECT

Extended Response

▸ **Is 7 more than 4?**

▸ **Name a number that is less than 20.**

▸ **What number is equal to 9?**

▸ **Describe a time when you have used the word** *equal* **outside of class.**

Encourage student discussion of these questions and answers.

Progress Monitoring

If... English learners are not proficient enough to describe their understanding of the word *more*,

▸ **Then...** use comprehension checks that do not rely heavily on language by having children use manipulatives or by using prompts such as *Show me thumbs up if you need more.*

4 ASSESS

Informal Assessment

Have students complete the activity below to make sure they understand the vocabulary. As students use each word:

1. Check understanding.

2. Correct errors.

3. Recheck for understanding.

- Show students two sets of objects. Have them describe the sets using the terms *more* and *less*.
- Have students define *compare* in their own words.

For each word, use the following rubric to assign a score.

The student can repeat the word when prompted. (1 point)

The student knows the word but does not know its meaning. (2 points)

The student has a vague idea of the word's meaning. (3 points)

The student knows the word and can use the word in context. (4 points)

More Counting and Comparing

Week 3

Objective
Students can compare and order numbers and match patterns and quantities using the terms *enough*, *too few*, and *too many*.

Vocabulary
- **compare** To think about how things are alike and different
- **enough** The correct amount
- **equal** Having the same amount
- **few** Made up of a small number
- **many** Made up of a large number
- **numeral** A symbol that represents a number

Materials

Program Materials
- Number Cards (1–10), p. 129
- Number Cards (11–20), p.130

Additional Materials
- crayons, red and blue
- dry, loose beans
- erasers
- paper clips
- pencil boxes or shoe boxes

1 WARM UP

Introduce each vocabulary word to students. Say the word aloud and have students repeat it.

Practice counting aloud from 1–20, both as a group and individually. If students are comfortable counting to 20, then practice counting to 30 and beyond.

Organize students into pairs. Distribute to each pair of students three piles of beans in different amounts between 1 and 20, a copy of Number Cards (1–10), p. 129, and a copy of Number Cards (11–20), p. 130. Students will work together to match the correct Number Card with each pile of beans.

2 ENGAGE

Place at least seven pencil boxes or shoe boxes on the floor. Explain that each pencil box needs to be filled with a pencil, a paper clip, a red crayon, a blue crayon, and an eraser.

Count the number of boxes aloud, and place the corresponding Number Card on the floor beside the boxes. Display different numbers of each item, counting aloud as a group as each set is displayed. Place the corresponding Number Card on the floor beside each set of items.

▶ **I want to put one pencil in each box. Do I have enough pencils to do this?** Answers should reflect the number of boxes and pencils.

▶ **Are there too many or too few pencils?** Answers should reflect the number of boxes and pencils.

Repeat these questions with different numbers of each item.

Display eight pencils, six erasers, five paper clips, eight red crayons, and five blue crayons.

▶ **How many pencils do I have?** eight

▶ **What other group is equal in number to the eight pencils?** red crayons

▶ **Compare the numbers of erasers and pencils. Are there more erasers or more pencils?** more pencils

▶ **How many more erasers do I need to have a number equal to the number of red crayons?** two more erasers

▶ **Are there enough boxes for each person in this room?** Answers should reflect the number of boxes and the number of people in the room.

Teacher Note

In this lesson, students use the term *enough* to compare number values. The *-ough* ending can be difficult to recognize, and therefore pronounce, by sight. Help English learners make a list of words that rhyme with *enough* and use the *-ough* ending, such as *tough* and *rough*. English learners may recognize the /uf/ sound in *enough* and should find additional words that use this sound but do not end in *-ough* such as *stuff, cuff,* and *fluff*. Encourage students to practice pronouncing the *-ough* ending.

Progress Monitoring

If... students need help determining how many more are needed, ▶ **Then...** help them count out each set to review the amounts.

3 REFLECT

Extended Response

► What does *enough* mean?

► How might you use *enough* in a sentence?

► How do we know if there are too many or too few of something?

Encourage student discussion of these questions and answers.

Progress Monitoring

| **If…** students are challenged with the concepts of *too many* or *too few*, | ▶ **Then…** work individually with students to describe scenarios that illustrate *too many* or *too few*. |

4 ASSESS

Informal Assessment

Have students complete the activity below to make sure they understand the vocabulary. As students use each word:

1. Check understanding.
2. Correct errors.
3. Recheck for understanding.

- Show students three sets of objects. Have students use the terms *many, few,* and *equal* to describe the sets of objects.
- Have students define *enough* in their own words.

For each word, use the following rubric to assign a score.

The student can repeat the word when prompted. (1 point)

The student knows the word but does not know its meaning. (2 points)

The student has a vague idea of the word's meaning. (3 points)

The student knows the word and can use the word in context. (4 points)

Matching Dot Sets to Numerals

Week 4

Objective
Students continue to compare and order numbers and can match Dot Set Cards to numerals.

Vocabulary
- **higher number/lower number** A higher number has more, a lower number has less
- **match** To find two things that are the same
- **numeral** A symbol that represents a number
- **symbol** A letter, number, or picture that stands for something else

Materials
Program Materials
- Number Cards (1–10), p. 129
- Number Cards (11–20), p. 130
- Dot Set Cards (1–10), p. 131

WARM UP

Introduce each vocabulary word to students. Say the word aloud and have students repeat it.

Count aloud from 1 to 20. Begin again, but stop partway, and ask the students to count on from there.

Count aloud backward from 20 to 1. Begin again, but stop partway, and ask the students to count back from there.

Organize students into groups of four. Have students count aloud by sets of five by having the first student count from 1 to 5, the second student count from 6 to 10, the third student count from 11 to 15, and the fourth student count from 16 to 20. Repeat this activity by counting backward.

2 ENGAGE

Organize students into pairs. Distribute a copy of Number Cards (1–10), p. 129, and Number Cards (11–20), p. 130, to each pair of students. Display a Dot Set Card or two Dot Set Cards that equal a number between 1 and 20, and explain to students that they will work in pairs to play a matching game with their Number Cards.

Ask students to name the number on the Dot Set Card and find the matching numeral in their Number Cards. Give each pair a point for correctly identifying and pronouncing each number. The pair with the most points at the end of the matching game wins.

To continue the game, ask questions about each numeral.

- ▶ **What number is one number higher than this numeral?**
- ▶ **What number is one number lower than this numeral?**
- ▶ **Compare these two Dot Set Cards. Which number is higher? Which is lower? Are they equal?**

Teacher Note

Some English learners might be confused by numbers described as *higher* or *lower* because these words commonly refer to position. Discuss with children that *higher* refers to numbers that come later in the counting sequence, while *lower* refers to numbers that come earlier in the sequence.

Progress Monitoring

If… students primarily recognize *match* in a different context,	▶ **Then…** encourage them to brainstorm items that go together, such as matching socks, a matching team uniform, a hat and mittens, and so on.

3 REFLECT

Extended Response

▶ What are the next five numbers after 10?

▶ What five numbers come next after I count 10, 9, 8, 7, 6?

▶ How do you know what number is being shown on a Dot Set Card?

▶ What does it mean to be a *match*?

Encourage student discussion of these questions and answers.

Progress Monitoring

If... students need an additional challenge,	▶ **Then...** encourage them to practice counting forward as high as they can as well as counting backward from increasingly larger numbers.

4 ASSESS

Informal Assessment

Have students complete the activity below to make sure they understand the vocabulary. As students use each word:

1. Check understanding.

2. Correct errors.

3. Recheck for understanding.

- Show students a set of numbers. Have students compare the numbers using the terms *higher number* and *lower number*.

- Have students describe counting on and counting back. Have students count on and count back from given numbers.

For each word, use the following rubric to assign a score.

The student can repeat the word when prompted. (1 point)

The student knows the word but does not know its meaning. (2 points)

The student has a vague idea of the word's meaning. (3 points)

The student knows the word and can use the word in context. (4 points)

Number Sequence and Number Lines

Week 5

Objective
Students can understand the meanings of the terms *before* and *after* and can compare and order numbers.

Vocabulary
- **after** Behind; to come later
- **before** In front of; to come first
- **far** Not close to; a long way away
- **near** Close to
- **line** A straight path that extends infinitely in opposite directions, thought of as having length, but no thickness

Materials

Program Materials
Vocabulary Card: *line*

Additional Materials
- masking tape
- a soft ball or bean bag that can be gently tossed
- self-sticking notes

1 WARM UP

Introduce each vocabulary word to students. Say the word aloud and have students repeat it.

Show students the *line* **Vocabulary Card** while saying the word aloud.

As a group, generate a conversation using the words *before* and *after*.

► **Do you put your socks on before or after your shoes?**

► **Do you eat dinner before or after you come home from school?**

► **Do you put away your toys before or after you play with them?**

► **Are you in second grade before or after first grade?**

Ask students to describe the events of their day using the terms *before* and *after*.

2 ENGAGE

Use masking tape to create a number line on the floor. Ask students to stand on each number so numbers decrease to the students' left and increase to the right. Draw a corresponding number line on the board.

Have students pass a soft ball or bean bag up and down the line, saying each number aloud sequentially. Mirror the movement of the ball on the number line on the board by attaching self-sticking notes on each number that is referenced.

► **If we begin passing the ball at 1, will it reach the person on 6 before or after it reaches the person on 5?** after

► **If we pass the ball to the person on 8, will (he or she) receive it before or after the person on 10?** before

Call out different numbers, and have students pass the ball up the number line to that number.

► **Start the ball on 1. Pass the ball to the person on 5. Pass it two more times. Where does it land?** 7

► **Start the ball on 1. Pass the ball to the person on 3. Pass it two more times. Pass it one more time. Where does it land?** 6

► **Is the person standing on 10 near or far from the person standing on 1?** far

► **Is the person standing on 3 near or far from the person standing on 2?** near

Demonstrate the ball's movement on the board by affixing self-sticking notes sequentially to the number line. Repeat this activity by passing the ball through different sequences of numbers.

Teacher Note

One strategy for helping English learners comprehend new math concepts is to use models and demonstrations. Using a number line is one way to demonstrate counting on and counting back. It provides a tool for beginning students to show their understanding by pointing to the answer or by moving a marker on the number line.

Progress Monitoring

If... students have difficulty recognizing the sequential progression up or down a number line,	► **Then...** offer them the opportunity to mirror the ball's progression by placing the self-sticking notes on the board.

3 REFLECT

Extended Response

▶ Is Monday before or after Tuesday?

▶ Is 9 before or after 8?

▶ What is the number that comes before 7?

▶ Is the number 4 near or far from the number 5?

▶ What helps you remember how to understand a number line?

▶ Can you think of anything outside the classroom that resembles a number line?

Encourage student discussion of these questions and answers.

Progress Monitoring

If... students can fluently identify and sequence numbers,	▶ Then... have them skip count as they pass the math ball up or down the number line.

4 ASSESS

Informal Assessment

Have students complete the activity below to make sure they understand the vocabulary. As students use each word:

1. Check understanding.

2. Correct errors.

3. Recheck for understanding.

- Show students a number. Have students define *after* and name numbers that come after the given number.

- Have students define *before* and name numbers that come before the given number.

- Have students describe the relationships of pairs of objects using the terms *near* and *far*.

For each word, use the following rubric to assign a score.

The student can repeat the word when prompted. (1 point)

The student knows the word but does not know its meaning. (2 points)

The student has a vague idea of the word's meaning. (3 points)

The student knows the word and can use the word in context. (4 points)

Final Oral Assessment

Administer the appropriate Final Oral Assessment, pp. 32–33, to each student. Use the rubric to determine students' levels of vocabulary acquisition.

Use the Student Assessment Record, page 142, to record the assessment results.

Final Oral Assessment 1, p. 32

Weeks 1–5 Final Oral Assessment 1 (Beginning English Learners)

Directions: Read each question to the student, and record his or her oral responses. Some questions have teacher directions. Teacher directions are indicated in italics. Allow students to use pencil and paper to work their responses.

1. I will **count forward.** Say *yes* if I have counted correctly or *no* if I have missed any numbers. *Count aloud from 1 to 10. Say each number.* **yes**

2. I will **count forward.** Say *yes* if I have counted correctly or *no* if I have missed any numbers. *Count aloud from 1 to 20. Skip the number 12.* **no**

3. I will **count forward.** Say *yes* if I have counted correctly or *no* if I have missed any numbers. *Count backward from 20 to 10. Skip the number 13.* **no**

4. Which pile has **more?** *Display two piles of paper clips, one with twelve paper clips and the other with five paper clips.* **Student should indicate the pile with twelve paper clips.**

5. How many **more** paper clips are in the larger pile? **seven more paper clips**

6. **Compare** these piles of paper clips. Are these piles **equal?** *Display two piles of paper clips, each with six paper clips.* **yes**

7. **Compare** these piles of paper clips. How many are in each pile? *Display three piles of paper clips: the first with four, the second with seven, and the third with ten.* **four, seven, ten**

8. Which pile of paper clips has the **most? Student should indicate the pile with ten paper clips.**

9. Which pile of paper clips has the **least? Student should indicate the pile with four paper clips.**

10. I want to give one paper clip to each of my friends. I have eight friends. Do I have **enough** paper clips? *Display six paper clips.* **no**

- **Minimal Understanding:** 0–3 of Questions 1–10 correct
- **Basic Understanding:** 4–7 of Questions 1–10 correct
- **Secure Understanding:** 8–10 of Questions 1–10 correct

Use the Student Assessment Record, page 142, to record the assessment results.

Weeks 1–5 Final Oral Assessment 2 (Intermediate and Advanced English Learners)

Directions: Read each question to the student, and record his or her oral responses. Some questions have teacher directions. Teacher directions are indicated in italics. Allow students to use pencil and paper to work their responses.

1. I will count aloud. **Count on** from where I stop. *1, 2, 3, 4, 5.* **6, 7, 8, 9, 10**

2. I will count aloud. **Count on** from where I stop. *11, 12, 13, 14, 15.* **16, 17, 18, 19, 20**

3. **Compare** these groups of paper clips. How many are in each group? Which group has the most? *Display three groups of paper clips: the first with eight, the second with thirteen, the third with eighteen.* **eight, thirteen, eighteen; The group with eighteen has the most.**

4. I want to give a piece of pizza to each of my friends. If my pizza has eight pieces and I have seven friends, do I have **enough** pizza? *Draw eight pieces of pizza and seven stick figures on a piece of paper.* **yes**

5. I want to give a present to each of my cousins. I have sixteen cousins and fifteen presents. Do I have **enough** presents? *Draw sixteen stick figures and fifteen squares.* **no**

6. How many **more** presents would I need if I have sixteen cousins and only fifteen presents? **one more present**

7. Do we put on our shoes **before** or **after** we put on our socks? **after**

8. What are some of the things you do **after** school? **Answers will vary.**

9. What number comes **after** 8? *Draw a number line with points labeled from 1 to 10 on a piece of paper.* **9**

10. If I draw **three** more dots, how many dots will I have? *Draw a number line with points labeled from 1 to 10 on a piece of paper. Draw solid dots over 1, 2, 3, and 4.* **seven**

- **Minimal Understanding:** 0–3 of Questions 1–10 correct
- **Basic Understanding:** 4–7 of Questions 1–10 correct
- **Secure Understanding:** 8–10 of Questions 1–10 correct

Use the Student Assessment Record, page 142, to record the assessment results.

Weeks 6–10

Section at a Glance

In this section, students will learn the vocabulary associated with **Number Worlds,** Level C, Weeks 6–10. Students use number lines to understand concepts of *before* and *after*, compare and order numbers, add and subtract whole numbers, and count up or down from a number.

Before beginning the section, assess students' general knowledge of math vocabulary using the Individual Oral Assessment on page 35.

How Students Learn Vocabulary

One strategy for helping English learners comprehend new math concepts is using models and demonstrations. Number lines provide English learners with physical evidence of a number sequence and assist with adding and subtracting whole numbers. Number lines also provide tangible opportunities to enact and understand concepts of *before* and *after*.

Academic Vocabulary Taught in Weeks 6–10

Week 6

add To combine numbers or put together numbers

after Behind; to come later

before In front of; to come first

minus sign A symbol that means "take away"

plus sign A symbol that means "add"

Week 7

after Behind; to come later

before In front of; to come first

between In the middle of two things

Week 8

after Behind; to come later

before In front of; to come first

between In the middle of two things

far Not close to; a long way away

near Close to

neighbor In math, the number before or after a certain number

Week 9

add Put together

altogether In all; total

plus sign A symbol that means "add"

less A smaller number or amount

more A larger number or amount

Week 10

add Put together

between In the middle of two things

equal Having the same amount

next The one after this one

plus sign A symbol that means "add"

Weeks 6–10 Individual Oral Assessment

Directions: Read each question to the student, and record his or her oral responses. Some questions have teacher directions. Teacher directions are indicated in italics. Allow students to use pencil and paper to work their responses.

1. Is 8 a **neighbor** of 9? **yes**

2. Does 8 come **after** 9? **no**

3. Is second grade **before** third grade? **yes**

4. Should I brush my teeth **before** or **after** I have eaten dinner? **after**

5. What number comes **after** 8? *Draw a number line with points labeled from 1 to 10.* **9**

6. What number comes **before** 11? *Draw a number line with points labeled from 1 to 20.* **10**

7. Is this a **plus sign**? *Write a plus sign on a piece of paper.* **yes**

8. Does a plus sign mean that you should **add** together or take away? **add together**

9. Does this mean to **add** one? *Write +1 on a piece of paper.* **yes**

10. How would you read this number sentence? Will the answer come **before** or **after** 3? *Write 3 + 1 on a piece of paper.* **3 + 1; after**

11. What is the number sentence for this picture? Will the answer come **before** or **after** 3? *Draw ... + .. = on a piece of paper.* **3 + 2 =; after**

12. What is the answer when you **add** 3 + 2? **5**

13. What is the answer when you **add** 7 + 1 = ? *Write 7 + 1 = on the board, and say it aloud.* **8**

- **Beginning English Learners:** 0–3 of Questions 1–10 correct
- **Intermediate English Learners:** 4–7 of Questions 1–10 correct
- **Advanced English Learners:** 8–10 of Questions 1–10 correct
- If the student is able to answer Questions 11–13, then he or she can understand the mathematics taught in this unit but may still have difficulty with the academic vocabulary.

Use the Student Assessment Record, page 142, to record the assessment results.

More Number Sequence and Number Lines

Week 6

Objective
Students continue addressing the concepts of *before* and *after* and can add and subtract whole numbers.

Vocabulary
- **add** To combine numbers or put together numbers
- **after** Behind; to come later
- **before** In front of; to come first
- **minus sign** A symbol that means "take away"
- **plus sign** A symbol that means "add"

Materials

Program Materials
- Vocabulary Cards: *add*
- Plus One, Minus One, Plus Two, Minus Two, p. 132
- Two-color counters (optional)

Additional Materials
- magnets for a number line on a board
- tape for a number line on the floor

1 WARM UP

Introduce each vocabulary word to students. Say the word aloud and have students repeat it.

Hold up the *add* **Vocabulary Card** and say the word aloud. Have students repeat the word.

Practice counting as a group from 1 to 20. After reviewing the week's vocabulary, address the words *before* and *after* by providing students with examples of the terms.

▶ **I will put a stamp on a letter before I mail it.**

▶ **I cross the street after I have looked both ways.**

Organize students into pairs. Have students create a list of three things they do before school and three things they do after school. Encourage students to share their lists with the class and practice using the vocabulary.

2 ENGAGE

Draw a number line on the board. Label the points from 1 to 20. Alternatively, you may create a number line on the floor with tape. Engage students with questions about the number line.

▶ **What number comes after 9?** 10

▶ **What number comes before 12?** 11

▶ **What number comes after 16?** 17

▶ **What does the plus sign mean?** to add together

▶ **Which way will my marker move on the number line when I add numbers?** toward the larger numbers or to the right

Write a number on the board, invite a student to the board, and ask the student to place a magnet on the number.

Then tape a Plus One or Minus One card next to the written number, and ask the student to move the magnet to the correct number. Then write the corresponding number sentence on the board.

▶ **Start on 5. Follow the instructions on this card to move your magnet to the correct number.**

Tape a Plus One card next to the written number 5.

▶ **What number will you move to?** 6

▶ **Describe this movement with the words *before* and *after*.** 6 is 1 space after 5, or 5 is one before 6

Repeat this activity with different +1 and −1 combinations. When students are comfortable with +1 and −1, introduce +2 and −2 combinations.

Teacher Note

Show English learners a plus sign. You might choose to have students make a plus sign with their fingers, pointing out that making a plus sign requires adding together their two hands, just as a plus sign means adding together numbers.

Progress Monitoring

If... students need additional practice moving markers up or down a number line,	▶ **Then...** help them enact the Plus One/Minus One movements on a number line on the floor by walking up or down the number line and describing their movements with the words *before* and *after*.

3 REFLECT

Extended Response

▶ How do you know that 8 comes after 7?

▶ How can you identify a plus sign?

▶ What helps you remember what a plus sign means?

Encourage student discussion of these questions and answers.

Progress Monitoring	
If... students need additional practice adding on a number line,	▶ **Then...** help them model addition equations on number lines with Counters.

4 ASSESS

Informal Assessment

Have students complete the activity below to make sure they understand the vocabulary. As students use each word:

1. Check understanding.

2. Correct errors.

3. Recheck for understanding.

- Have students identify a plus sign and a minus sign.

- Show students a number line. Place a magnet on the number line. Have students identify numbers that come before and after the number indicated.

For each word, use the following rubric to assign a score.

The student can repeat the word when prompted. (1 point)

The student knows the word but does not know its meaning. (2 points)

The student has a vague idea of the word's meaning. (3 points)

The student knows the word and can use the word in context. (4 points)

Number Neighborhoods

Week 7

Objective
Students can understand the concepts of *before, after,* and *between* and can compare and order numbers on a number line.

Vocabulary
- **after** Behind; to come later
- **before** In front of; to come first
- **between** In the middle of two things

Materials

Program Materials
Plus One, Minus One, Plus Two, Minus Two, p. 132

Additional Materials
- bean bag or small, soft ball to toss
- dry beans
- glue
- index cards
- string
- tape for a number line on the floor

1 WARM UP

Introduce each vocabulary word to students. Say the word aloud and have students repeat it.

Practice counting aloud from 1 to 20 by passing a soft bean bag around the room. If a student misses a number or says an incorrect number, ask students to find the correct number. Return to 1, and repeat the number sequence again. If students need an extra challenge, pass the bean bag and count aloud backward from 20 to 1.

2 ENGAGE

Create a number line with points labeled from 1 to 20 on the floor with tape. Invite a student to stand on the number 11, and ask the following questions.

▶ **What number are you standing on?** 11
▶ **What is the number before your number?** 10
▶ **What is the number after your number?** 12
▶ **What number will you stand on when I show you the +1 card?** 12
▶ **What number will you stand on when I show you the +2 card?** 13
▶ **What number will you stand on when I show you the −1 card?** 10

Ask two students to stand on the number line, one on 7 and one on 9. Ask each student the following questions.

▶ **What is the number before your number?** *6; 8*
▶ **What is the number after your number?** *8; 10*
▶ **What is the number between your numbers?** *8*

Continue asking students about their places on the number line, and have them describe movement up and down a number line using the words *before, after,* and *between*.

Teacher Note

Number twenty index cards from 1 to 20. Write *1* on the first card, *2* on the second card, and so on. Glue the same number of small objects, such as dry beans, to each card. For example, glue two beans on the 2 card and five beans on the 5 card. Punch a hole in each card, and hook it with a paper clip. Clip the paper clip to a string to create a number line that can be manipulated. Have students take turns hanging the number cards in sequence on the number line.

Progress Monitoring

If... students are confident with the new vocabulary,	▶ Then... create a number line using larger numbers and continue asking students to define movement up and down the number line.

3 REFLECT

Extended Response

▶ **What are the next two numbers after 7?**

▶ **What are the next two numbers before 13?**

▶ **Describe where you sit in the classroom using the word *between*.**

▶ **Use the words *before, after,* and *between* to describe the number 14.**

Encourage student discussion of these questions and answers.

Progress Monitoring

If... English learners are challenged by the /tw/ sound in *between*,	▶	**Then...** use sound examples such as *tweet* and *twice* to help students practice the sound.

4 ASSESS

Informal Assessment

Have students complete the activity below to make sure they understand the vocabulary. As students use each word:

1. Check understanding.

2. Correct errors.

3. Recheck for understanding.

Show students two numbers on a number line. Have students describe the relationship of the given numbers using the terms *after, before,* and *between*.

For each word, use the following rubric to assign a score.

The student can repeat the word when prompted. (1 point)

The student knows the word but does not know its meaning. (2 points)

The student has a vague idea of the word's meaning. (3 points)

The student knows the word and can use the word in context. (4 points)

More Number Neighborhoods

Week 8

Objective
Students can use the spatial terms *before* and *after* with the number sequence and can order numbers.

Vocabulary
- **after** Behind; to come later
- **before** In front of; to come first
- **between** In the middle of two things
- **far** Not close to; a long way away
- **near** Close to
- **neighbor** In math, the number before or after a certain number

Materials
Program Materials
- Number Cards (1–10), p. 129
- Number Cards (11–20), p.130
- Plus One, Minus One, Plus Two, Minus Two, p. 132

Additional Materials
- baskets
- tape for hopscotch path on the floor

1 WARM UP

Introduce each vocabulary word to students. Say the word aloud and have students repeat it.

Organize students into two teams. Distribute a copy of Number Cards (1–10), p. 129, and Number Cards (11–20), p. 130, to each student. Write two sequential numbers between 1 and 20 on two pieces of paper, and have one student from each team find the neighboring numbers in their Number Cards and place the correct cards on either side of the written numbers.

For example, write *12* and *13* on two pieces of paper. Both students should place the 11 card on the left and the 14 card on the right of the paper to complete the sequence. Ask both students to read their answers aloud. Each student receives one point for a correct answer. The team with the most points at the end of the **Warm Up** wins.

2 ENGAGE

Create a hopscotch path with squares from 1 to 20 on the floor using tape. Make sure the squares are big enough for students to stand in them. Fill one basket with Number Cards and another basket with Plus One, Minus One, Plus Two, Minus Two cards. Invite a student to choose a card from the first basket.

▶ **Begin at the start of the hopscotch path, and hop to the number written on your card.**

Pick a +1, −1, +2, or −2 card from the other basket.

▶ **Hop up or down the hopscotch path to the correct number.**

▶ **Describe your hopping. Which number did you start on? What are the neighbors of that number? Is your number near the start or far away from the start? Did you hop ahead or back? Is the number you landed on before or after the number you started on?**

▶ **What is the number sentence that describes your hopping?**

While one student hops the path, the other students will write the corresponding number sentence on their pieces of paper. Repeat the activity for each student.

Progress Monitoring

If... students are concentrating more on the hopping than the counting,	▶ Then... have students move a bean bag or small ball up and down the hopscotch path.

3 REFLECT

Extended Response

► What does *between* mean?

► If I start on 5 and hop ahead two spaces, where will I land?

► What are the neighbors of that number?

► What numbers come next after 15?

Encourage student discussion of these questions and answers.

Progress Monitoring

If... students need a review before playing the hopscotch game,	► **Then...** draw a number line on the board and practice counting aloud together to remind students of the number sequence.

4 ASSESS

Informal Assessment

Have students complete the activity below to make sure they understand the vocabulary. As students use each word:

1. Check understanding.

2. Correct errors.

3. Recheck for understanding.

- Show students a number line, and point to a number. Have students name the neighbors of the given number.

- Have students describe the relationship of two numbers using the terms *before, after, between, near,* and *far.*

For each word, use the following rubric to assign a score.

The student can repeat the word when prompted. (1 point)

The student knows the word but does not know its meaning. (2 points)

The student has a vague idea of the word's meaning. (3 points)

The student knows the word and can use the word in context. (4 points)

Week 9

Objective

Students can understand the meanings of the terms *plus sign* and *add* and can add whole numbers.

Vocabulary

- **add** Put together
- **altogether** In all; total
- **less** A smaller number or amount
- **more** A larger number or amount
- **plus sign** A symbol that means "add"

Materials

Program Materials
- Vocabulary Card: *add*
- Number Cards (1–10), p. 129
- Number Cards (11–20), p.130
- Plus One, Minus One, Plus Two, Minus Two, p. 132

Additional Materials
index cards

1 WARM UP

Introduce each vocabulary word to students. Say the word aloud and have students repeat it.

Show students the *add* **Vocabulary Card** and review the concept of addition.

▶ **I will draw a number story on the board, and you will take turns saying the number story aloud.**

Draw ★ + ★★★ = on the board.

▶ **How many stars are there altogether?** 4

▶ Draw ★ ★★★ after the = on the board.

▶ **What is the correct number sentence for this drawing?** $1 + 3 = 4$

Repeat this activity with different addition sentences.

2 ENGAGE

Ask students to draw two number lines on a piece of paper. Have students label the points from 1 to 20.

Organize students into pairs. Distribute a copy of Number Cards (1–10), p. 129, Number Cards (11–20), p. 130, and Plus One, Minus One, Plus Two, Minus Two, p. 132, to each pair of students.

▶ **Work in pairs to create addition number stories using your +1 and +2 Cards, your Number Cards, and a number line.**

▶ **One of you will use your Number Cards and a +1 Card or +2 Card to create a number sentence, and the other will solve the number sentence by drawing it on a number line.**

▶ **When you have each written two addition number stories and drawn two addition stories on number lines, please share your work with the class. Use our new vocabulary words in your description of your work.**

Teacher Note

The *pl* blend in *plus* may be difficult for some English learners to pronounce. Brainstorm sounds and words that use the *pl* blend, and help students be creative with their ideas. *Plop* and *plink* are some fun examples. It is important for students to feel confident when practicing a new sound. This is more effectively achieved when students can make their efforts into a game.

Progress Monitoring

If... students are confident with +1 and +2 addition sentences,

▶ **Then...** encourage them to write *+3* and *+0* on index cards and create addition sentences with these new addends.

3 REFLECT

Extended Response

▶ **What is this sign? Display a plus sign.**

▶ **What does *add* mean?**

▶ **Describe a time when you have added something outside of class.**

▶ **I want to add 3 and 4. How many will there be altogether? How would you say this number sentence?**

Encourage student discussion of these questions and answers.

Progress Monitoring

If... some English learners understand the math concepts but are struggling with describing their understanding, ▶	**Then...** pair them with English learners who are more proficient with the language.

4 ASSESS

Informal Assessment

Have students complete the activity below to make sure they understand the vocabulary. As students use each word:

1. Check understanding.

2. Correct errors.

3. Recheck for understanding.

- Have students identify a plus sign.
- Have students define *add* in their own words.

For each word, use the following rubric to assign a score.

The student can repeat the word when prompted. (1 point)

The student knows the word but does not know its meaning. (2 points)

The student has a vague idea of the word's meaning. (3 points)

The student knows the word and can use the word in context. (4 points)

More Adding

Week 10

Objective

Students continue to compare and order numbers, can understand the meanings of the terms *plus sign* and *add,* and can add whole numbers.

Vocabulary

- **add** Put together
- **between** In the middle of two things
- **equal** Having the same amount
- **next** The one after this one
- **plus sign** A symbol that means "add"

Materials

Program Materials
- Vocabulary Card: *add*
- Number Cards (1–10), p. 129

Additional Materials
- dry, loose beans
- paper cups
- tape

1 WARM UP

Introduce each vocabulary word to students. Say the word aloud and have students repeat it.

Show students the *add* **Vocabulary Card** and review the concept of addition.

Draw a number line on the board. Label the points from 1 to 20. Organize students into pairs or small groups. Have students work in groups to describe a number. Each group will use the following four sentences about their number:

► **Our number is one more than _____.**

► **The next number is would be _____.**

► **Our number is between _____ and _____.**

Have groups share their descriptions with the class.

2 ENGAGE

Organize the class into two groups, and distribute a paper cup and a handful of beans to each student. Distribute Number Cards (1–10) to students. Ask them to model their numbers with beans and place the beans in the cup.

Write _____ + _____ = _____ on the board.

Call a number from each group. The students with those numbers will come to the front of the room, tape their Number Cards or write their numbers in the correct blank, and add their beans to find a total. As a class, count aloud the beans and write the equation while saying it aloud.

► **From Group A, I would like the person with number 6 to come to the front of the room. From Group B, I would like the person with number 2 to come to the front of the room. Have students place their numbers on the board.**

► **Our equation now looks like 6 + 2 = _____.**

Have students say the equation aloud.

► **Let's add the two groups of beans to find our total. Count aloud as a group.**

Repeat this activity with different combinations of addends, and help students say the equations aloud.

Teacher Note

Students may want to play this game in small groups and create addition equations of their own. Allow them to do this and keep track of their equations with pencil and paper.

Progress Monitoring

If... students are confident with smaller addend equations,	▶	**Then...** encourage them to create larger addition equations by adding larger addends.

3 REFLECT

Extended Response

▶ What is the next number after 12?

▶ When I add 4 and 3, will my answer be before or after 4?

▶ How would you tell this number story? Draw eight stars and three stars.

▶ What helps you remember the meaning of a plus sign?

Encourage student discussion of these questions and answers.

Progress Monitoring

If… you think students need additional practice with the lesson vocabulary,	▶ **Then…** help them make a vocabulary chart with written definitions and drawn examples.

4 ASSESS

Informal Assessment

Have students complete the activity below to make sure they understand the vocabulary. As students use each word:

1. Check understanding.
2. Correct errors.
3. Recheck for understanding.

- Have students define *add* in their own words and solve an addition problem.
- Have students identify a plus sign.

For each word, use the following rubric to assign a score.

The student can repeat the word when prompted. (1 point)

The student knows the word but does not know its meaning. (2 points)

The student has a vague idea of the word's meaning. (3 points)

The student knows the word and can use the word in context. (4 points)

Final Oral Assessment

Administer the appropriate Final Oral Assessment, pp. 46–47, to each student. Use the rubric to determine students' levels of vocabulary acquisition.

Use the Student Assessment Record, page 142, to record the assessment results.

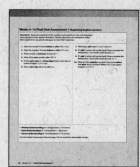

Final Oral Assessment 1, p. 46

Weeks 6–10 Final Oral Assessment 1 (Beginning English Learners)

Directions: Read each question to the student, and record his or her oral responses. Some questions have teacher directions. Teacher directions are indicated in italics. Allow students to use pencil and paper to work their responses.

1. Does the number 9 come **before** or **after** 10? **before**

2. Does the number 18 come **before** or **after** 17? **after**

3. What number is **between** 10 and 12? **11**

4. What is the **next** number after 19? **20**

5. Is this a **plus sign** or a **minus sign**? *Draw a plus sign on a piece of paper.* **a plus sign**

6. Does a **plus sign** tell us to add? **yes**

7. What does **add** mean? **to put together**

8. If I **add** 1, where will my dot land? *Draw a number line labeled from 1 to 10 with a solid dot over 5.* **6**

9. If I **add** 2, where will my dot land? *Draw a number line labeled from 1 to 20 with a solid dot over 13.* **15**

10. What are the **neighbor** numbers that come **before** and **after** these numbers? *Write 9, 10, 11 on a piece of paper.* **8 and 12**

- **Minimal Understanding:** 0–3 of Questions 1–10 correct
- **Basic Understanding:** 4–7 of Questions 1–10 correct
- **Secure Understanding:** 8–10 of Questions 1–10 correct

Use the Student Assessment Record, page 142, to record the assessment results.

Weeks 6–10 Final Oral Assessment 2 (Intermediate and Advanced English Learners)

Directions: Read each question to the student, and record his or her oral responses. Some questions have teacher directions. Teacher directions are indicated in italics. Allow students to use pencil and paper to work their responses.

1. What are the **neighbor** numbers that come **before** and **after** these numbers? *Write 12, 13, 14 on a piece of paper.* **11 and 15**

2. What does a **plus sign** mean? *Draw a plus sign on a piece of paper.* **to add or put together**

3. If my dot starts on 11 and I **add** 1, where will my dot land? *Draw a number line labeled from 1 to 20 on a piece of paper with a solid dot over 11.* **12**

4. Use the word *plus* to say a number sentence for the previous question. $11 + 1 = 12$

5. What is the number story for this drawing? What does $5 + 1$ **equal?** *Draw five stars and one star.* $5 + 1 = 6$

6. What is the number story for this drawing? What does $4 + 2$ **equal?** *Draw four stars and two stars.* $4 + 2 = 6$

7. What is the number story for this drawing? What does $6 + 2$ **equal?** *Draw six stars and two stars.* $6 + 2 = 8$

8. What is the number story for this drawing? What does $7 + 3$ **equal?** *Draw seven stars and three stars.* $7 + 3 = 10$

9. Use circles to draw a picture for this addition equation. What numbers did you **add?** *Write $6 + 3 = 9$ on a piece of paper.* **Student should draw six circles + three circles = nine circles; 6 and 3.**

10. Solve the addition equation $9 + 3 = $ _____. What numbers does the answer fall **between?** **11 and 13**

- **Minimal Understanding:** 0–3 of Questions 1–10 correct
- **Basic Understanding:** 4–7 of Questions 1–10 correct
- **Secure Understanding:** 8–10 of Questions 1–10 correct

Use the Student Assessment Record, page 142, to record the assessment results.

Weeks 11–15

Section at a Glance

In this section, students will learn the vocabulary associated with **Number Worlds,** Level C, Weeks 11–15. Students are expected to understand the meaning and value of *zero*, add and subtract whole numbers, write and solve equations, compare and order numbers, begin learning about graphs, and understand the meaning of the word *equality*. Before beginning the section, assess students' general knowledge of math vocabulary using the Individual Oral Assessment on page 49.

How Students Learn Vocabulary

English learners will learn new vocabulary by associating objects and actions, such as adding or subtracting counters, with corresponding words and concepts.

Academic Vocabulary Taught in Weeks 11–15

Week 11

add To combine numbers or put together numbers

equal Having the same amount; identical in value or notation

plus sign A symbol that means "add"

symbol A letter, number, or picture that has special meaning or stands for something else

zero The number that, when used as an addend, leaves any number unchanged

Week 12

add To combine numbers or put together numbers

equal Having the same amount; identical in value or notation

equal sign A symbol that means "having the same amount"

equation A number story

minus sign A symbol that means "take away"

plus sign A symbol that means "add"

subtract To take away, as in one quantity from another

Week 13

add To combine numbers or put together numbers

after Behind; to come later

before In front of; to come first

far Not close to; a long way away

near Close to

plus sign A symbol that means "to add"

Week 14

equal Having the same amount; identical in value or notation

equal sign A symbol that means "having the same amount"

equation A number story

plus sign A symbol that means "add"

solve To figure out; to find the answer

symbol A letter, number, or picture that has special meaning or stands for something else

Week 15

compare To think about how things are the same and different

equal Having the same amount; identical in value or notation

equation Number story

graph A visual representation of data

higher number A number with a greater value when counting up

thermometer A tool used to measure how hot or cold something is

Weeks 11–15 Individual Oral Assessment

Directions: Read each question to the student, and record his or her oral responses. Some questions have teacher directions. Teacher directions are indicated in italics. Allow students to use pencil and paper to work their responses.

1. Is this a **plus sign?** *Draw a plus sign on a piece of paper.* **yes**

2. Is this an **equal sign?** *Draw an equal sign on a piece of paper.* **yes**

3. Does an **equal sign** mean "having the same amount"? **yes**

4. I will count aloud. **Count on** from where I stop. 1, 2, 3, 4, 5, 6 **7, 8, 9, 10**

5. I will count aloud. **Count on** from where I stop. 11, 12, 13, 14, 15 **16, 17, 18, 19, 20**

6. Do we use a **minus sign** when we **add** or when we **subtract? when we subtract**

7. Does *zero* mean "none" or "many"? **none**

8. If I have five pencils in my left hand and eight pencils in my right hand, do I have an **equal** amount in each hand? **no**

9. If I want to combine, or put together, both sets of pencils, would I **add** or **subtract** them? **add the pencils**

10. Which number is **higher,** 8 or 13? **13**

11. What does this number story, or **equation,** say? *Write $5 + 4 = 9$ on a piece of paper.* $5 + 4 = 9$

12. How would you describe this **equation?** *Draw ★★★ + ★★★★ = ★★★★★★★ on a piece of paper.* $3 + 4 = 7$

13. **Solve** this **equation,** and read it aloud. *Write $5 - 2 = $ _____ on a piece of paper.* $5 - 2 = 3$

- **Beginning English Learners:** 0–3 of Questions 1–10 correct
- **Intermediate English Learners:** 4–7 of Questions 1–10 correct
- **Advanced English Learners:** 8–10 of Questions 1–10 correct
- If the student is able to answer Questions 11–13, then he or she can understand the mathematics taught in this unit but may still have difficulty with the academic vocabulary.

Use the Student Assessment Record, page 142, to record the assessment results.

Sequencing Numbers

Week 11

Objective
Students can add and subtract whole numbers and can understand the meaning of the term *zero*.

Vocabulary
- **add** To combine numbers or put together numbers
- **equal** Having the same amount; identical in value or notation
- **plus sign** A symbol that means "add"
- **symbol** A letter, number, or picture that has special meaning or stands for something else
- **zero** The number that, when used as an addend, leaves any number unchanged

Materials

Program Materials
- Vocabulary Cards: *add, equal, zero*
- Two-color counters

Additional Materials
coffee can or other opaque container

1 WARM UP

Introduce each vocabulary word to students. Say the word aloud and have students repeat it.

Hold up the *add, equal,* and *zero* **Vocabulary Cards.** Say each word aloud, and have students repeat the word.

Tell students that to play today's game they should listen very carefully to see if they can catch you making a counting mistake. Model counting up to 10, and then invite volunteers to count to 10 alone.

Ask students to listen carefully while you count. Have students raise their hands if they hear you make a mistake.

▶ **1, 2, 3, 4, 5, 6, 8, 9, 10** Award one tally point to students if they correctly tell you which number was left out; award one tally point to yourself if students do not catch the mistake.

Repeat this procedure, counting from 1 to 10, sometimes leaving out different numbers or saying the same number twice and other times counting correctly. When students are ready, start counting from 10 to 20.

2 ENGAGE

Explain to students that this activity will help them learn about the number zero. Practice saying the word *zero* aloud, and remind students of its definition. Introduce the symbol for zero (**0**). Ask students to translate *zero* to their primary language and discuss how the value of zero, or nothing, remains the same in every language.

Distribute at least five counters to each student.

▶ **Pass around the can. Place one counter in it while saying the number sentence aloud.**

▶ **When we begin, the can will be empty, or have nothing in it. What number of counters will this represent?** zero

▶ **I will pass the empty can to the first person, and he or she will say this addition equation: $0 + 1 = 1$.**

▶ **The first person will then put a counter into the can. The first person will pass the can to the second person, who will say this addition equation: $1 + 1 = 2$.**

▶ **The second person will then put a counter in the can.**

Repeat the passing until the equation reaches $4 + 1 = 5$. Then have students pass the can around and subtract one counter each time, saying the equation aloud until the answer is zero.

As students become more comfortable with this game, increase the number of counters so students create $+2$ and $+3$ equations. Students may need to pour the counters onto a desk and count them to find the total.

Teacher Note

English learners may need practice recognizing and pronouncing the /z/ in *zero*. Offer them creative opportunities to use this sound by saying *Zoom! Zap! Zing!* and other fun noises with the letter *z*. When the practice exercise is engaging and entertaining, students will probably feel less self-conscious about incorrect pronunciation or having to make repeated efforts.

Progress Monitoring

If... students have trouble determining the value of zero,	▶ Then... help them recognize zero in the number sequence by drawing a number line.

3 REFLECT

Extended Response

► What is the value of zero?

► What number is larger than zero and smaller than 2?

► If I add zero to 13, how many will I have?

► Describe a time you gave away all of something and had zero left.

Encourage student discussion of these questions and answers.

Progress Monitoring

If... students need further help understanding the value of zero,	▶ Then... include it in your counting activities throughout the day. For example, before students line up for recess, note that there are zero students in line, then have one student stand in line, and so on.

4 ASSESS

Informal Assessment

Have students complete the activity below to make sure they understand the vocabulary. As students use each word:

1. Check understanding.

2. Correct errors.

3. Recheck for understanding.

Draw a zero on the board. Have students identify the number and define it.

For each word, use the following rubric to assign a score.

The student can repeat the word when prompted. (1 point)

The student knows the word but does not know its meaning. (2 points)

The student has a vague idea of the word's meaning. (3 points)

The student knows the word and can use the word in context. (4 points)

Writing Equations

Week 12

Objective
Students can write equations and can solve equations with more than two addends.

Vocabulary
- **add** To combine numbers or put together numbers
- **equal** Having the same amount; identical in value or notation
- **equal sign** A symbol that means "having the same amount"
- **equation** A number story
- **minus sign** A symbol that means "take away"
- **plus sign** A symbol that means "add"
- **subtract** To take away, as in one quantity from another

Materials

Program Materials
- Vocabulary Cards: *add, equal, subtract*
- Planet Equations, p. 133

Additional Materials
- index cards
- paper, with planet names, pictures, and equations

1 WARM UP

Introduce each vocabulary word to students. Say the word aloud and have students repeat it.

After defining the vocabulary words, show students the *add, equal,* and *subtract* **Vocabulary Cards** while saying each word aloud.

Invite the class to practice identifying and pronouncing equation symbols by playing a group flash card game. Draw a plus sign, a minus sign, and an equal sign on three index cards. Flash an equation symbol index card at students, and encourage them to call aloud the symbol you have displayed. If some students do not participate, have students raise their hands when you display a card, and call on individual students to identify the sign.

2 ENGAGE

Post pieces of paper with the name of a planet, a drawing of that planet, and a number story described below. Distribute a copy of Planet Equations, p. 133, to each student. Have students visit each "planet" to solve the addition or subtraction equation it represents.

Find Mercury. Write the equation in numbers:

★★ + ★ + ★★★ = ★★★★★★ $2 + 1 + 3 = 6$

Find Venus. Write the equation in numbers:

★★★ + ★ + ★★★★ = ★★★★★★★★
$3 + 1 + 4 = 8$

Find Earth. Write the equation in numbers:

★★★★★ − ★★★ = ★★ $5 - 3 = 2$

Find Mars. Write the equation in numbers:

★★★★★★ − _____ = ★★★★★★ $6 - 0 = 6$

Find Jupiter. Draw the number story:

$2 + 3 + 2 = 7$ OO + OOO + OO = OOOOOOO

Find Saturn. Draw the number story:

$5 - 4 = 1$ OOOOO − OOOO = O

Find Uranus. Draw the number story:

$3 + 0 + 2 = 5$ OOO + _____ + OO = OOOOO

Find Neptune. Write the equation in numbers:

★★★ + ★★ + ★★★★ = ★★★★★★★★★
$3 + 2 + 4 = 9$

▶ **Travel to each planet and write the equation or drawing shown on that planet. Write your answer beside the planet name on your worksheet.**

Invite students to share their answers with the rest of the class by describing its equations using the terms *plus, minus,* and *equals*.

Teacher Note

This activity can be done with students in pairs or small groups if they are not yet comfortable with equations.

Progress Monitoring

If... students are not yet familiar with planets or planet names,	▶	Then... change the planet names to state names or names of objects, and reference as many or as few equations as you deem necessary.

3 REFLECT

Extended Response

▶ What does a plus sign look like?

▶ What does a minus sign look like?

▶ What helps you remember how an equal sign looks?

▶ If I want to write a number story about putting groups of numbers together, what symbols, or signs, would I use?

Encourage student discussion of these questions and answers.

Progress Monitoring

If... students have trouble adding three or more addends at a time,	▶ **Then...** engage students in a sequential counting activity in which three students each have a small number of counters and the first two students add their counters by counting aloud sequentially. The third student continues counting aloud to complete the addition equation.

4 ASSESS

Informal Assessment

Have students complete the activity below to make sure they understand the vocabulary. As students use each word:

1. Check understanding.

2. Correct errors.

3. Recheck for understanding.

- Show students a plus sign, a minus sign, and an equal sign. Have students identify each sign.

- Have students define *equation* in their own words and explain how the word *equation* is related to the word *equals*.

For each word, use the following rubric to assign a score.

The student can repeat the word when prompted. (1 point)

The student knows the word but does not know its meaning. (2 points)

The student has a vague idea of the word's meaning. (3 points)

The student knows the word and can use the word in context. (4 points)

Counting and Adding

Week 13

Objective
Students can understand the meanings of the terms *near, far, before,* and *after* by sequencing numbers on a number line and solving equations.

Vocabulary
- **add** To combine numbers or put together numbers
- **after** Behind; to come later
- **before** In front of; to come first
- **far** Not close to; a long way away
- **near** Close to
- **plus sign** A symbol that means "to add"

Materials
Program Materials
Vocabulary Cards: *before, after*

1 WARM UP

Introduce each vocabulary word to students. Say the word aloud and have students repeat it.

Show students the *before* and *after* **Vocabulary Cards** while saying each word aloud.

Have students count as a group and as individuals from 1 to 20. Then engage students to count as a group from 1 to 30. Ask students if they know what number comes after 39. If they correctly identify 40, practice counting aloud from 1 to 40 and ask students if they recognize a number pattern in the groups of ten.

2 ENGAGE

Draw a number line with points labeled from 1 to 50 on the board. Also draw a two-column chart on the board, labeled *Students* and *Teacher*.

▶ **We are going to play a game. I will ask questions and call on individual students to answer me. If your answer is correct, the students will receive one point. If your answer is incorrect, I will get a point. At the end of the game, whoever has the most points wins.**

▶ **Circle 42. Is this number near 50 or near 0?** near 50

▶ **Circle 18. What is the number before 18?** 17

▶ **Circle 27. Is this number far from 30?** no

▶ **Circle 34. What is the next number after 34?** 35

▶ **Circle 19. Write 25 and 39 on the board. Which number is nearer to 19, 25 or 39?** 25

▶ **Circle 10. Write 10 + 4 on the board. What direction does this number story, or equation, tell me to move on the number line?** to the right or toward the larger numbers

▶ **Circle 20. I want to move to 26. In a number story, or equation, would I use a plus sign or a minus sign to show this move?** plus sign

▶ **Circle 15. I want to move to 13. In a number story, or equation, would I use a plus sign or a minus sign to show this move?** minus sign

Continue to ask different questions to engage students in discussion of movement on a number line. You may want to incorporate equations that indicate progression on a number line.

Teacher Note

This activity can be played with any number line most appropriately suited for the class. You might decide to focus on a specific section of the number sequence to fit the needs of the students.

Progress Monitoring

If... students need smaller number sequences,	▶ Then... create shorter number lines, increasing their range as students grow more comfortable with identifying progression on a number line.

3 REFLECT

Extended Response

► What are some numbers that are near 5 on the number line?

► What are some numbers that are far from 5 on the number line?

► What does a plus sign mean?

► What does a plus sign in an equation on a number line mean?

Encourage student discussion of these questions and answers.

Progress Monitoring

If... students need additional reference for *near, far, before,* or *after* for a number sequence,	► Then... invite them to practice the vocabulary by comparing objects in the classroom with these spatial terms.

4 ASSESS

Informal Assessment

Have students complete the activity below to make sure they understand the vocabulary. As students use each word:

1. Check understanding.

2. Correct errors.

3. Recheck for understanding.

- Have students define *far* and *near* in their own words and describe the relationships of objects using those terms.

- Have students define *before* and *after* in their own words and describe the relationships of numbers in a sequence using those terms.

For each word, use the following rubric to assign a score.

The student can repeat the word when prompted. (1 point)

The student knows the word but does not know its meaning. (2 points)

The student has a vague idea of the word's meaning. (3 points)

The student knows the word and can use the word in context. (4 points)

Making Equations

Week 14

Objective

Students can understand the meanings of the terms *equality* and *equal* and can write addition equations.

Vocabulary

- **equal** Having the same amount; identical in value or notation
- **equal sign** A symbol that means "having the same amount"
- **equation** A number story
- **plus sign** A symbol that means "add"
- **solve** To figure out; to find the answer
- **symbol** A letter, number, or picture that has special meaning or stands for something else

Materials

Program Materials
- Vocabulary Card: *equal*
- Addition for 10, p. 134

Additional Materials
blue and red crayons

1 WARM UP

Introduce each vocabulary word to students. Say the word aloud and have students repeat it.

Show students the *equal* **Vocabulary Card** while saying the word aloud. Discuss the meaning of equal.

▶ **I will draw a number story on the board, and you will take turns saying the number story aloud. Then we must solve each problem by figuring out the answer.**

Draw 2 blue circles and 4 red circles on a large sheet of paper.

▶ **What is the correct number sentence for this drawing?** $2 + 4 =$

▶ **What number will solve this equation?** 6

Continue by drawing models and solving the following equations.

$3 + 1 + 4 = 8$ $10 - 9 = 1$
$5 + 2 = 7$ $1 + 4 + 5 = 10$
$9 - 0 = 9$ $12 - 12 = 0$

Repeat this activity with different addition and subtraction equations. Help students practice identifying and saying *plus, minus,* and *equal*.

2 ENGAGE

Distribute a blue crayon, a red crayon, and a copy of Addition for 10, p. 134, to each student. Organize students into pairs.

▶ **Work with your partner to complete this worksheet. Use your blue and red crayons to create addition equations that equal 10.**

▶ **You will create a different equation for each row of circles. Try to use the number zero at least once.**

▶ **Share your equations with the class by writing them on the board with numbers and saying them aloud.**

Teacher Note

This worksheet is easily created for equations of any number. If students need to practice the activity with smaller numbers, create a worksheet for equations of 5 or 6. If students are already confident with addends of 10, then create a worksheet with addends of higher numbers.

Progress Monitoring

If... students need visual assistance with the Engage activity,	▶ Then... draw a number line on the board or encourage students to draw their own number lines for reference.

③ REFLECT

Extended Response

▶ What symbol do we use to show that two things are the same?

▶ What is another word for *number story*?

▶ How do you know what number goes to the right of an equal sign?

▶ How can you check your answer in an equation?

▶ Can you think of other times at home or at school when you add?

Encourage student discussion of these questions and answers.

Progress Monitoring	
If... students need an additional challenge,	▶ **Then...** encourage them to work in pairs as one student says an addition equation aloud and the other writes the equation to practice recognizing addition vocabulary without visual reference.

④ ASSESS

Informal Assessment

Have students complete the activity below to make sure they understand the vocabulary. As students use each word:

1. Check understanding.
2. Correct errors.
3. Recheck for understanding.

- Have students explain how the terms *equal, equal sign,* and *equation* are related.
- Have students use lesson vocabulary to describe how they solve an equation.

For each word, use the following rubric to assign a score.

The student can repeat the word when prompted. (1 point)

The student knows the word but does not know its meaning. (2 points)

The student has a vague idea of the word's meaning. (3 points)

The student knows the word and can use the word in context. (4 points)

Graphing and Comparing Numbers

Week 15

Objective
Students can understand the meaning of the term *graph* and can describe and compare values depicted on a graph.

Vocabulary
- **compare** To think about how things are the same and different
- **equal** Having the same amount; identical in value or notation
- **equation** Number story
- **graph** A visual representation of data
- **higher number** A number with a greater value when counting up
- **thermometer** A tool used to measure how hot or cold something is

Materials
Program Materials
Vocabulary Card: *equal*

Additional Materials
- 4 pencils
- 5 paper clips
- 2 crayons

WARM UP

Introduce each vocabulary word to students. Say the word aloud and have students repeat it.

Show students the *equal* **Vocabulary Card** while saying the word aloud.

Draw a thermometer on the board. Label degrees from 0 to 20. Ask students if they recognize this shape. Discuss what a thermometer is and how it is used. Ask students to count on from 0 to 20 and count back from 20 to 0. Point to the numbers on the thermometer as they count aloud.

▶ **What is the highest number on this thermometer?** 20

▶ **What is the lowest number on this thermometer?** 0

▶ **Which number is higher than 13, 15 or 12?** 15

▶ **What number comes immediately before 16?** 15

▶ **If the thermometer were taller, what number would come after 20?** 21

2 ENGAGE

Explain that you will work as a group to create a graph. Draw a three-column, five-row graph on the board, and tape a pencil, a paper clip, and a crayon under the three columns. Number the rows 1 to 5, from bottom to top.

Display the remaining three pencils, four paper clips, and one crayon, and ask students to count aloud to determine how many there are of each item. Ask students to help you decide how to fill in the spaces on the graph.

▶ **How many spaces should we color for the pencils?** three

▶ **How many spaces should we color for the crayons?** one

▶ **How many spaces should we color for the paper clips?** four

▶ **Which set has the most? How do you know?** paper clips; it has the highest number when I count them

▶ **Which set has the least? How do you know?** crayons; it has the lowest number when I count them

▶ **What does it mean to compare?** to think about how things are the same and different

▶ **What do you compare with this graph?** the number of objects

▶ **Which set has the higher number, pencils or paper clips?** paper clips

Repeat this activity by creating graphs with different information and asking questions that use the week's vocabulary.

Teacher Note

English learners may need practice pronouncing the /f/ sound in *graph*. Help them generate a list of words that use this sound with *f* or *–ff*, such as *giraffe*, *stuff*, *fall*, and *February*. Then help them generate a list of words that use *–ph* to create the /f/ sound, such as *telephone*, *dolphin*, and *photo*.

Progress Monitoring

If... students can compare quantities on the graph fluently,	▶ **Then...** have them work with larger numbers.

3 REFLECT

Extended Response

► **How can graphs help us count and compare numbers?**

► **How do you know when two sets on a graph show the same amount?**

► **How does a graph show us the highest amount?**

Encourage student discussion of these questions and answers.

Progress Monitoring

If... students need an additional challenge,	► **Then...** have them create their own three-column graph depicting the items in their desk or book bag and share their findings with the class.

4 ASSESS

Informal Assessment

Have students complete the activity below to make sure they understand the vocabulary. As students use each word:

1. Check understanding.

2. Correct errors.

3. Recheck for understanding.

- Have students define *graph* in their own words and describe how they can use a graph to display data.

- Have students identify a thermometer and explain how it is used. Have students explain how a thermometer can be used to compare temperatures.

For each word, use the following rubric to assign a score.

The student can repeat the word when prompted. (1 point)

The student knows the word but does not know its meaning. (2 points)

The student has a vague idea of the word's meaning. (3 points)

The student knows the word and can use the word in context. (4 points)

Final Oral Assessment

Administer the appropriate Final Oral Assessment, pp. 60–61, to each student. Use the rubric to determine students' levels of vocabulary acquisition.

Use the Student Assessment Record, page 142, to record the assessment results.

Final Oral Assessment 1, p. 60

Weeks 11–15 Final Oral Assessment 1 (Beginning English Learners)

Directions: Read each question to the student, and record his or her oral responses.
Some questions have teacher directions. Teacher directions are indicated in italics.
Allow students to use pencil and paper to work their responses.

1. I will count aloud. **Count on** from where I stop. 6, 7, 8, 9, 10 **11, 12, 13, 14, 15**

2. What does a **plus sign** tell us to do? **add**

3. What does *add* mean? **to combine or put together**

4. What does a **minus sign** tell us to do? **to subtract, or take away**

5. What does an **equal sign** mean? **having the same amount**

6. What does *zero* mean? **none**

7. If I **add** 0 to 4, how many will I have? **4**

8. If I **subtract** 0 from 9, how many will I have? **9**

9. Solve this addition **equation.** *Draw a number line with points labeled from 0 to 10, and write* $2 + 5 =$ _____ *on a piece of paper.* $2 + 5 = 7$

10. Solve this subtraction **equation**. *Draw a number line with points labeled from 0 to 10, and write* $9 - 2 =$ _____ *on a piece of paper.* $9 - 2 = 7$

- **Minimal Understanding:** 0–3 of Questions 1–10 correct
- **Basic Understanding:** 4–7 of Questions 1–10 correct
- **Secure Understanding:** 8–10 of Questions 1–10 correct

Use the Student Assessment Record, page 142, to record the assessment results.

Weeks 11–15 Final Oral Assessment 2 (Intermediate and Advanced English Learners)

Directions: Read each question to the student, and record his or her oral responses. Some questions have teacher directions. Teacher directions are indicated in italics. Allow students to use pencil and paper to work their responses.

1. What is the **highest** number shown? *Draw a number line labeled from 0 to 10.* **10**

2. Which number is **higher,** 18 or 16? *Draw a number line labeled from 0 to 20.* **18**

3. Is this number **near** 20 or **near** 5? *Draw a number line labeled from 0 to 20, and circle 17.* **near 20**

4. **Solve** this **equation,** and read your answer aloud. *Draw a number line labeled from 0 to 20. Write* $0 + 3 = $ _____ *on a piece of paper.* $0 + 3 = 3$

5. **Solve** this **equation,** and read your answer aloud. *Draw a number line labeled from 0 to 20. Write* $9 + 4 = $ _____ *on a piece of paper.* $9 + 4 = 13$

6. **Solve** this **equation,** and read your answer aloud. *Draw a number line labeled from 0 to 20. Write* $15 - 2 = $ _____ *on a piece of paper.* $15 - 2 = 13$

7. Is this a **graph?** *Draw and label a three-column, five-row graph on a piece of paper. Draw a three-unit red bar, a two-unit yellow bar, and a five-unit blue bar.* **yes**

8. Which color has the **highest number?** *Reference the graph in Question 7.* **blue**

9. Write and **solve** this addition **equation.** *Draw three stars, two stars, and two stars on a piece of paper.* $3 + 2 + 2 = 7$

10. Write and **solve** this subtraction **equation.** *Draw*

★★★★★★★★★★ − ★★★★★★ = _____

on a piece of paper. $10 - 6 = 4$

- **Minimal Understanding:** 0–3 of Questions 1–10 correct
- **Basic Understanding:** 4–7 of Questions 1–10 correct
- **Secure Understanding:** 8–10 of Questions 1–10 correct

Use the Student Assessment Record, page 142, to record the assessment results.

Introduction

Weeks 16–20

Section at a Glance

In this section, students will learn the vocabulary associated with **Number Worlds,** Level C, Weeks 16–20. Students are expected to understand and use math symbols, understand the meaning of *equality*, and add and subtract whole numbers. Before beginning the section, assess students' general knowledge of math vocabulary using the Individual Oral Assessment on page 63.

How Students Learn Vocabulary

English learners will gain understanding of new vocabulary by relating it to familiar concepts outside the classroom. Students will recognize and use the terms *plus sign*, *minus sign*, and *equal sign* by understanding that they are symbols, much like a stop sign and an exit sign are symbols, that tell us what to do. Number lines and counters provide tools for relating addition and subtraction concepts to the vocabulary in this section.

Academic Vocabulary Taught in Weeks 16–20

Week 16

add To combine numbers or put together numbers

altogether In all; total

equal sign A symbol that means "having the same amount"

equation A number story

figure out To think about a problem and solve it; to find the answer

plus sign A symbol that means "add"

Week 17

after Behind; to come later

before In front of; to come first

between In the middle of two things

equal sign A symbol that means "having the same amount"

equation A number story

figure out To think about a problem and solve it; to discover the answer

plus sign A symbol that means "add"

Week 18

add To combine numbers or put together numbers

after Behind; coming later

before In front of; coming first

between In the middle of two things

minus sign A symbol that means "takeaway"

subtract To take away, as in one quantity from another

Week 19

count back To begin with a number and count backward, subtracting one each time

equal sign A symbol that means "having the same amount"

minus sign A symbol that means "take away"

number line A line with evenly spaced sequential numbers or units; a number line can be used to solve math problems

solve To figure out; to find the answer

subtract To take away, as in one quantity from another

Week 20

add To combine numbers or put together numbers

equal Having the same amount; identical in value or notation

equal sign A symbol that means "having the same amount"

equation A number story that includes an equal sign

minus sign A symbol that means "take away"

missing number The number needed to complete an equation

plus sign A symbol that means "add"

subtract To take away, as in one quantity from another

Weeks 16–20 Individual Oral Assessment

Directions: Read each question to the student, and record his or her oral responses. Some questions have teacher directions. Teacher directions are indicated in italics. Allow students to use pencil and paper to work their responses.

1. Is this a **plus sign**? *Draw a plus sign on a piece of paper.* **yes**

2. Is this a **minus sign**? *Draw a minus sign on a piece of paper.* **yes**

3. Does a **plus sign** mean "add"? **yes**

4. Does a **minus sign** mean "equal"? **no**

5. Does a **minus sign** mean "subtract"? **yes**

6. What does an **equal sign** mean when it is **between** numbers? *Draw an equal sign on a piece of paper.* **It means that the numbers are equal, or the same.**

7. I will **count back** aloud. Continue counting when I stop. *Count back from 10 to 5.* **5, 4, 3, 2, 1**

8. I will **count back** aloud. Continue counting when I stop. *Count back from 20 to 10.* **10, 9, 8, 7, 6, 5, 4, 3, 2, 1**

9. Is $4 + 6$ **equal** to 11? **no**

10. Is $5 - 3$ **equal** to 2? **yes**

11. Solve this **equation.** You may draw a picture. *Write $2 + 3 =$ _____ on a piece of paper.* **5**

12. Solve this **equation.** You may draw a picture. *Write $5 - 1 =$ _____ on a piece of paper.* **4**

13. Solve this **equation.** You may draw a picture. *Write $6 - 2 =$ _____ on a piece of paper.* **4**

- **Beginning English Learners:** 0–3 of Questions 1–10 correct
- **Intermediate English Learners:** 4–7 of Questions 1–10 correct
- **Advanced English Learners:** 8–10 of Questions 1–10 correct
- If the student is able to answer Questions 11–13, then he or she can understand the mathematics taught in this unit but may still have difficulty with the academic vocabulary.

Use the Student Assessment Record, page 142, to record the assessment results.

More Counting and Adding

Week 16

Objective
Students can review the meanings of the terms *plus sign, minus sign, equal sign,* and *equation* while recognizing the number sequence by counting backward.

Vocabulary
- **add** To combine numbers or put together numbers
- **altogether** In all; total
- **equal sign** A symbol that means "having the same amount"
- **equation** A number story
- **figure out** To think about a problem and solve it; to find the answer
- **plus sign** A symbol that means "add"

Materials
Program Materials
- Vocabulary Cards: *add, equal, subtract*
- Two-color counters

Additional Materials
index cards

1 WARM UP

Introduce each vocabulary word to students. Say the word aloud and have students repeat it.

Hold up the *add, equal,* and *subtract* **Vocabulary Cards.** Read each word aloud, and have students repeat the word.

▶ **Count forward with me from 1 to 20.**

▶ **Count back with me from 20 to 1.**

Count aloud from 1 to 10 and then from 1 to 20 as a group. Model the sequential counting by placing one counter on the table for each number. Distribute twenty counters to each student, and work together to model different number values with counters. Count aloud as you do so.

When you count back as a group, have students remove one counter each time they say a number.

▶ **In this lesson, we are going to use the plus sign and the equal sign in equations.**

▶ **What is a sign?** a letter, number, or picture that has special meaning or stands for something else

▶ **What are some signs we see outside of class? What do they mean?** Answers will vary.

2 ENGAGE

Distribute two index cards to each student. Ask students to draw a plus sign on one and an equal sign on the other. On a desk, make a pile of two counters, a pile of four counters, and a pile of six counters.

▶ **I will use my plus sign and equal sign to make an equation with these counters.**

▶ **Where should I put my plus sign?** between 2 and 4

▶ **Where should I put my equal sign?** between 4 and 6

▶ **My plus sign and equal sign make this equation: 2 + 4 = 6. Is this correct?** yes

▶ **I will make different equations with counters on the desk. You will each have a turn to place your plus sign and equal sign on the desk to make an equation. Then you will read the equation aloud.**

Create different equations with counters on the desk, and invite students to take turns creating and saying the equations. Ask the following questions for each equation.

▶ **How did you figure that out?**

▶ **How many do you have altogether?**

Teacher Note

Students may want to create their own equations once they are comfortable with this group game. Allow them to work in pairs to create an equation with piles of counters and equation symbols.

Progress Monitoring

If... students need an extra challenge,	▶ Then... write addition equations on the board without the visual reference of counters.

3 REFLECT

Extended Response

▶ **What number comes next when we count back 10, 9, 8?**

▶ **What do signs do?**

▶ **How do you know when to use an equal sign?**

Encourage student discussion of these questions and answers.

Progress Monitoring

| **If...** students need practice with a vocabulary word or particular sound, | ▶ **Then...** encourage them to find words that have similar sounds in both English and their primary language. |

4 ASSESS

Informal Assessment

Have students complete the activity below to make sure they understand the vocabulary. As students use each word:

1. Check understanding.
2. Correct errors.
3. Recheck for understanding.

- Have students identify the plus sign and the equal sign and explain how they are used.
- Have students define *altogether* in their own words and describe an addition equation using the word.

For each word, use the following rubric to assign a score.

The student can repeat the word when prompted. (1 point)

The student knows the word but does not know its meaning. (2 points)

The student has a vague idea of the word's meaning. (3 points)

The student knows the word and can use the word in context. (4 points)

Solving Equations

Week 17

Objective
Students can understand the meaning of the term *equal*, can compare and order numbers, and can use math symbols to add whole numbers.

Vocabulary
- **add** Behind; to come later
- **before** In front of; to come first
- **between** In the middle of two things
- **equal sign** A symbol that means "having the same amount"
- **equation** A number story
- **figure out** To think about a problem and solve it; to discover the answer
- **plus sign** A symbol that means "add"

Materials

Program Materials
Vocabulary Cards: *add, equal, subtract*

Additional Materials
- dot cubes
- tape to make hopscotch path

1 WARM UP

Introduce each vocabulary word to students. Say the word aloud and have students repeat it.

After defining the vocabulary words, show students the *add, equal,* and *subtract* **Vocabulary Cards.** Say each word aloud.

As a group, play "What Number Am I?" Present students with mystery problems. Sample problems include the following.

- ▶ **I come two numbers after 5. I come one number before 8. I am one more than 6. What number am I?** 7
- ▶ **I am more than 10. I am less than 15. The number right before me is 11. The number right after me is 13. What number am I?** 12
- ▶ **I am less than 20. I am more than 18. What number am I?** 19
- ▶ **I am one group of 10 and 4 ones. I am more than 12, and I am more than 13. I am less than 16, and I am less than 15. What number am I?** 14

2 ENGAGE

With masking tape, create a hopscotch path labeled from 1 to 20 on the floor. Make sure students have a pencil and paper for this activity.

- ▶ **I will throw two dot cubes, and we will create an addition equation from those numbers. I will write the equation on the board while you write it on your paper.**

Suppose you rolled three dots and five dots.

- ▶ **How should we read this equation?** $3 + 5 = 8$
- ▶ **What is the answer?** 8
- ▶ **Who would like to hop this equation on the hopscotch path?**
- ▶ **Where will you begin?** on 3; accept "at the start"
- ▶ **Where will you land when we have added both numbers?** 8
- ▶ **What number comes before your number?** 7
- ▶ **What number comes after your number?** 9
- ▶ **What numbers are you between?** 7 and 9

As students grow comfortable with two-addend equations, advance to equations with three addends generated by three dot cubes. Remind students to write the equations as they play the game, and call on volunteers to read each problem aloud.

Teacher Note ✏
It may help English learners to have a visual framework from which they answer the equations. Write _____ + _____ + _____ = _____ on the board to provide a sentence frame for their answers.

Progress Monitoring

If... students need additional practice adding three-addend equations,	▶ **Then...** help them read the equations aloud and break them into pieces by adding the first two addends together and then adding the third addend for the total.

3 REFLECT

Extended Response

▶ Why do we use a plus sign when we hop forward on the hopscotch path?

▶ How would you describe this game to someone new in our class?

▶ What number is between 10 and 12?

▶ How do you figure out equations with more than two numbers?

Encourage student discussion of these questions and answers.

Progress Monitoring

If... students need additional practice with this activity or adding with three addends,	▶	**Then...** distribute dot cubes and a number line on which students can draw and solve equations.

4 ASSESS

Informal Assessment

Have students complete the activity below to make sure they understand the vocabulary. As students use each word:

1. Check understanding.
2. Correct errors.
3. Recheck for understanding.

- Show students a number line, and indicate two numbers. Have students identify numbers that come before the given numbers, numbers that come after the given numbers, and numbers between the given numbers.

- Have students draw a plus sign and an equal sign and explain how each sign is used.

For each word, use the following rubric to assign a score.

The student can repeat the word when prompted. (1 point)

The student knows the word but does not know its meaning. (2 points)

The student has a vague idea of the word's meaning. (3 points)

The student knows the word and can use the word in context. (4 points)

Adding and Subtracting

Week 18

Objective
Students can use spatial terms to describe the number sequence and use addition and subtraction in successive operations.

Vocabulary
- **add** To combine numbers or put together numbers
- **after** Behind; coming later
- **before** In front of; coming first
- **between** In the middle of two things
- **minus sign** A symbol that means "take away"
- **subtract** To take away, as in one quantity from another

Materials
Program Materials
- Vocabulary Card: *add, equal, subtract*
- Number Cards (1–10), p. 129
- Number Cards (11–20), p. 130
- Grocery Store, p. 135
- Two-color counters

1 WARM UP

Introduce each vocabulary word to students. Say the word aloud and have students repeat it.

After defining the vocabulary words, show students the *add, equal,* and *subtract* **Vocabulary Cards** while saying each word aloud.

Use masking tape to create a hopscotch path from 1 to 20 on the floor. Invite students to draw a Number Card and hop to that number on the path.

▶ **What numbers are you between?**

▶ **What number comes before you?**

▶ **What number is two spaces after you?**

2 ENGAGE

Distribute ten counters and a copy of Grocery Store, p. 135, to each student. Tell students that they are going on a pretend grocery shopping trip to buy food, and they have ten counters to spend.

Students should decide which items they can buy and still have at least one counter left over. Have students take turns telling the group what they will buy, one purchase at a time, adding and subtracting when necessary.

Ask questions that lead children to explain the processes they applied to determine what they could purchase.

▶ **Did you add together prices until you were close to the ten-counter limit and then subtract?**

▶ **Did you subtract each item from 10 as you chose it?**

Teacher Note

This activity usually is a popular one with students and provides substantial practice in addition and subtraction. You may find it useful to create a poster-sized version of the grocery guide to display in front of the class and to use as a focal point for whole-class discussions.

Progress Monitoring

If... you think students are comfortable articulating equations to 10,	▶ **Then...** increase the number of counters they can use to "purchase" groceries.

3 | REFLECT

Extended Response

▶ What does *before* mean?

▶ What does *subtract* mean?

▶ Describe a recent shopping trip you experienced.

Encourage student discussion of these questions and answers.

Progress Monitoring	
If... students understand the math concepts but struggle with the English language component of the lesson,	▶ **Then...** pair them with partners who have more advanced English comprehension.

4 | ASSESS

Informal Assessment

Have students complete the activity below to make sure they understand the vocabulary. As students use each word:

1. Check understanding.

2. Correct errors.

3. Recheck for understanding.

- Have students explain how they used addition and subtraction in the game. Encourage students to use lesson vocabulary in their explanations.

- Have students define *add* and *subtract* in their own words.

For each word, use the following rubric to assign a score.

The student can repeat the word when prompted. (1 point)

The student knows the word but does not know its meaning. (2 points)

The student has a vague idea of the word's meaning. (3 points)

The student knows the word and can use the word in context. (4 points)

Week 19

Objective

Students can reference the number sequence by counting backward and subtract small quantities from single- and double-digit numbers.

Vocabulary

- **count back** To begin with a number and count backward, subtracting one each time
- **equal sign** A symbol that means "having the same amount"
- **minus sign** A symbol that means "take away"
- **number line** A line with evenly spaced sequential numbers or units; a number line can be used to solve math problems
- **solve** To figure out; to find the answer
- **subtract** To take away, as in one quantity from another

Materials

Program Materials
- Vocabulary Card: *add, equal, subtract*
- Number Lines (1–20), p. 136
- Two-color counters

Additional Materials
- tape for a hopscotch path or number line on the floor
- 20 magnets or self-sticking notes

1 WARM UP

Introduce each vocabulary word to students. Say the word aloud and have students repeat it.

Display the *add, equal,* and *subtract* **Vocabulary Cards.** Say the words aloud.

Use masking tape to create a hopscotch path or number line labeled from 1 to 20 on the floor. Invite individual students to hop up and down the path as the class counts aloud from 1 to 20 and back from 20 to 1.

2 ENGAGE

Distribute a copy of Number Lines (1–20), p. 136, and twenty counters to each student. Draw a number line with points labeled from 1 to 20 on the board. Explain to students that they will use the number lines to solve subtraction equations.

▶ **We will work as a group to solve this first subtraction equation.**

Write $12 - 6 =$ _____ on the board.

▶ **Who would like to read this equation aloud?**

Place magnets or self-sticking notes above numbers 1 to 12.

▶ **I want to subtract 6 from 12, so I will take away six magnets.**

Remove six magnets, counting aloud as you do so.

▶ **How many magnets are left? six**

▶ **So my equation is $12 - 6 = 6$.**

▶ **I will write another equation on the board, and you will use counters at your desks to solve the equation.**

Write different subtraction equations for students to solve. Give each student an opportunity to model and say a subtraction equation aloud at the board.

Teacher Note

Some English learners may feel more confident displaying and articulating an equation in front of a small group instead of the entire class. If needed, encourage students to work in small groups on this activity.

Progress Monitoring

If... students can fluently count up and back in the number sequence,	▶	**Then...** challenge them to hop a subtraction equation on a number line.

3 | REFLECT

Extended Response

▶ How can you check your answer when you subtract?

▶ What helps you determine an answer in subtraction?

▶ Can you think of other times at school or at home when you subtract?

Encourage student discussion of these questions and answers.

Progress Monitoring	
If... students struggle with subtraction from double-digit numbers,	▶ **Then...** practice subtracting 1 from higher numbers to create comfort with larger numbers and remind students that subtracting is like counting back.

4 | ASSESS

Informal Assessment

Have students complete the activity below to make sure they understand the vocabulary. As students use each word:

1. Check understanding.

2. Correct errors.

3. Recheck for understanding.

• Have students describe how to count back on a number line to solve a subtraction problem.

Remind students to use lesson vocabulary in their descriptions.

• Write a subtraction problem on the board. Have students use a number line to solve the problem.

For each word, use the following rubric to assign a score.

The student can repeat the word when prompted. (1 point)

The student knows the word but does not know its meaning. (2 points)

The student has a vague idea of the word's meaning. (3 points)

The student knows the word and can use the word in context. (4 points)

Subtracting and Predicting

Week 20

Objective

Students can understand different ways to say *subtract*, can add and subtract whole numbers, and continue recognizing and using math symbols.

Vocabulary

- **add** To combine numbers or put together numbers
- **equal** Having the same amount; identical in value or notation
- **equal sign** A symbol that means "having the same amount"
- **equation** A number story that includes an equal sign
- **minus sign** A symbol that means "take away"
- **missing number** The number needed to complete an equation
- **plus sign** A symbol that means "add"
- **subtract** To take away, as in one quantity from another

Materials

Program Materials
- Vocabulary Card: *add, equal, subtract*
- Number Lines (1–20), p. 136 (optional)

Additional Materials
- index cards
- small ball (optional)

1 WARM UP

Introduce each vocabulary word to students. Say the word aloud and have students repeat it.

Show students the *add, equal,* and *subtract* **Vocabulary Cards** while saying the words aloud.

The variety of words used to describe subtraction, such as *take away, minus, less, subtracted from,* and so on, may be puzzling for English learners. Help students brainstorm a list of words and phrases used to describe subtraction.

2 ENGAGE

Write these symbols and words on index cards: =, +, −, *add, subtract, minus, equal, solve,* and *equation*.

Explain to students that they will work as a group to win more points than the teacher by answering questions correctly.

▶ **I will show you a card and call on one of you to tell me what card I have displayed. If you are correct, the students will earn a point. If you are incorrect, I will earn a point.**

▶ **Once we have played the game with cards, I will write equations on the board, and you will find the missing number. If you have the correct answer when I call on you, the students will earn a point.**

Create various addition and subtraction equations with one- and two-digit numbers. Encourage students to work with pencil and paper to find the answers.

Teacher Note

English learners may need practice with the /s/ sound in *missing*. Encourage them to hiss like a snake or make the sound of air escaping a ball to create the *ss* sound. Help them generate a list of words that use the *ss* sound, such as *miss, kiss, hiss,* and so on.

Progress Monitoring

If... students need help subtracting,	▶ **Then...** distribute copies of Number Lines (1–20), p. 136, and help them map equations on the worksheet.

3 REFLECT

Extended Response

▶ **What is another way to say *minus*?**

▶ **How can you check your answer in a subtraction equation?**

▶ **What do you like better, adding or subtracting? Why?**

Encourage student discussion of these questions and answers.

Progress Monitoring	
If... students can fluently identify and sequence numbers,	▶ **Then...** have them skip count as they pass a math ball up or down the number line.

4 ASSESS

Informal Assessment

Have students complete the activity below to make sure they understand the vocabulary. As students use each word:

1. Check understanding.
2. Correct errors.
3. Recheck for understanding.

- Have students describe a missing number and explain how they solve for a missing number in an addition problem.
- Have students describe how they solve for a missing number in a subtraction problem.

For each word, use the following rubric to assign a score.

The student can repeat the word when prompted. (1 point)

The student knows the word but does not know its meaning. (2 points)

The student has a vague idea of the word's meaning. (3 points)

The student knows the word and can use the word in context. (4 points)

Final Oral Assessment

Administer the appropriate Final Oral Assessment, pp. 74–75, to each student. Use the rubric to determine students' levels of vocabulary acquisition.

Use the Student Assessment Record, page 142, to record the assessment results.

Final Oral Assessment 1, p. 74

Weeks 16–20 Final Oral Assessment 1 (Beginning English Learners)

Directions: Read each question to the student, and record his or her oral responses. Some questions have teacher directions. Teacher directions are indicated in italics. Allow students to use pencil and paper to work their responses.

1. I will **count back.** Did I count correctly? *Count aloud from 20 to 1.* **yes**

2. I will **count back.** Continue counting when I stop. *Count aloud 10, 9, 8, 7.* **6, 5, 4, 3, 2, 1**

3. Does 14 come **before** 15? **yes**

4. Does 18 come **after** 19? **no**

5. What number is **between** 5 and 7? **6**

6. Is this a **plus sign?** *Write a plus sign on a piece of paper.* **yes**

7. Is this a **minus sign?** *Write a minus sign on a piece of paper.* **yes**

8. **Solve** this **equation.** *Write* $8 + 5 =$ _____ *on a piece of paper.* **13**

9. **Solve** this **equation.** *Write* $3 + 1 + 5 =$ _____ *on a piece of paper.* **9**

10. **Solve** this **equation.** *Write* $5 - 4 =$ _____ *on a piece of paper.* **1**

- **Minimal Understanding:** 0–3 of Questions 1–10 correct
- **Basic Understanding:** 4–7 of Questions 1–10 correct
- **Secure Understanding:** 8–10 of Questions 1–10 correct

Use the Student Assessment Record, page 142, to record the assessment results.

Weeks 16–20 Final Oral Assessment 2 (Intermediate and Advanced English Learners)

Directions: Read each question to the student, and record his or her oral responses. Some questions have teacher directions. Teacher directions are indicated in italics. Allow students to use pencil and paper to work their responses.

1. What number comes **after** 14? *Draw a number line labeled from 1 to 20 on a piece of paper. Circle 14.* **15**

2. **Between** what numbers is 8? *Point to the number line again, and circle 8.* **7 and 9**

3. If a carrot costs three counters and you have ten counters, how many counters will you have left after you buy a carrot? Do you **add** or **subtract** to **solve** the problem? **Seven counters; subtract**

4. How do you write and say the **equation** used for buying the carrot? Use math signs in your answer. **10 − 3 = 7**

5. What does this word say? *Write **solve** on a piece of paper.* **solve**

6. What does *solve* mean? **to find the answer**

7. Find the **missing number.** *Write* 3 + _____ = 5 *on a piece of paper.* **2**

8. **Solve** this **equation.** *Write* 7 − 2 = _____ *on a piece of paper.* **5**

9. **Solve** this **equation.** *Write* 4 + 5 + 1 = _____ *on a piece of paper.* **10**

10. **Solve** this **equation.** *Write* 17 − 7 = _____ *on a piece of paper.* **17 − 7 = 10**

- **Minimal Understanding:** 0–3 of Questions 1–10 correct
- **Basic Understanding:** 4–7 of Questions 1–10 correct
- **Secure Understanding:** 8–10 of Questions 1–10 correct

Use the Student Assessment Record, page 142, to record the assessment results.

Weeks 21–25

Section at a Glance

In this section, students will learn the vocabulary associated with *Number Worlds,* Level C, Weeks 21–25. Students are expected to compare and order numbers, add and subtract whole numbers, understand and use math symbols, create and solve word problems, and identify numbers to 100. Before beginning the section, assess students' general knowledge of math vocabulary using the Individual Oral Assessment on page 77.

How Students Learn Vocabulary

Repetition is a key element in English learners' comprehension of math vocabulary. The lessons in this unit repeat words from prior lessons such as *add, subtract, minus sign, plus sign, equal,* and *equation.* This repetition will help students find these words familiar and therefore develop confidence with their mathematical understanding.

Academic Vocabulary Taught in Weeks 21–25

Week 21

add To combine numbers or put together numbers

compare To think about how things are alike and how they are different

equal Having the same amount; identical in value or notation

sum The answer to an addition problem

Week 22

add To combine numbers or put together numbers

equal Having the same amount; identical in value or notation

equal sign A symbol that means "having the same amount"

equation A number story

minus sign A symbol that means "take away"

pattern Something that repeats the same way each time

subtract To take away, as in one quantity from another

zero None; the number that, when used as an addend, leave any number unchanged

Week 23

add To combine numbers or put together numbers

equal Having the same amount; identical in value or notation

equal sign A symbol that means "having the same amount"

minus sign A symbol that means "take away"

missing number The number needed to complete an equation

plus sign A symbol that means "put together"

subtract To take away, as in one quantity from another

Week 24

missing number The number needed to complete an equation

pattern Something that repeats the same way each time

position Where something is; the location of something in relationship to other things

Week 25

after Behind; coming later

before In front of; coming first

far Not close to; a long way away

near Close to

ones column The numeral on the far right of a number that tells how many ones are in the number

tens column The numeral to the left of the ones column that tells how many tens are in a number

Weeks 21–25 Individual Oral Assessment

Directions: Read each question to the student, and record his or her oral responses. Some questions have teacher directions. Teacher directions are indicated in italics. Allow students to use pencil and paper to work their responses.

1. Is the number in the **ones column** a 3? *Write 23 on the board.* **yes**

2. Is the number in the **tens column** a 2? *Write 32 on the board.* **no**

3. When you add **zero** to another number, does it change the value of the number? **no**

4. What are these numbers? **Compare** these numbers. Which number is larger? *Draw a dot set card of 3 and a dot set card of 5 on a piece of paper.* **3 and 5; 5**

5. **Compare** these numbers. Which number is larger? *Write 9 and 19 on a piece of paper.* **19**

6. Is this the number **zero**? *Write 0 on a piece of paper.* **yes**

7. Does **zero** mean "many" or "none"? **none**

8. What does **add** mean? **to combine numbers or put together numbers**

9. What does *subtract* mean? **to take away, as in one quantity from another**

10. What math sign should I use in an addition **equation?** **plus sign**

11. Solve this **equation.** *Write* $14 + 4 =$ _____ *on a piece of paper.* $14 + 4 = 18$

12. **Solve** this **equation.** *Write* $8 + 9 =$ _____ *on a piece of paper.* $8 + 9 = 17$

13. **Solve** this **equation.** *Write* $22 - 0 =$ _____ *on a piece of paper.* $22 - 0 = 22$

- **Beginning English Learners:** 0–3 of Questions 1–10 correct
- **Intermediate English Learners:** 4–7 of Questions 1–10 correct
- **Advanced English Learners:** 8–10 of Questions 1–10 correct
- If the student is able to answer Questions 11–13, then he or she can understand the mathematics taught in this unit but may still have difficulty with the academic vocabulary.

Use the Student Assessment Record, page 142, to record the assessment results.

Adding and Comparing

Week 21

Objective
Students can compare and order whole numbers, add whole numbers, and understand math symbols.

Vocabulary
- **add** To combine numbers or put together numbers
- **compare** To think about how things are alike and how they are different
- **equal** Having the same amount; identical in value or notation
- **sum** The answer to an addition problem

Materials

Program Materials	Additional Materials
Vocabulary Cards: *add, equal*	dot cubes, 2 per pair

1 WARM UP

Introduce each vocabulary word to students. Say the word aloud and have students repeat it.

Hold up the *add* and *equal* **Vocabulary Cards.** Read each word aloud and have students repeat it.

As a group, create a chart of comparative and superlative forms of words, such as the following.

- high/higher/highest
- low/lower/lowest
- big/bigger/biggest
- small/smaller/smallest

2 ENGAGE

Organize students into groups of four with two teams of two students per group. Distribute two dot cubes to each pair of students.

► **With your partner, roll your two cubes. Add the cubes together for a sum.**

► **What number do you have? Compare the sums of the two teams.**

► **Which team has the highest number?**

► **How do you know it is the highest?**

► **The pair of students with the highest number gets a point.**

Continue playing the game as time permits. The team with the most points wins.

Teacher Note

Students may want to play the game with three dot cubes for an added challenge.

Progress Monitoring

If... students need assistance keeping track of their sums and points,	► **Then...** ask a fifth student to join the group to monitor the points earned.

3 REFLECT

Extended Response

▶ Which number is higher, 29 or 28?

▶ How can you check your answer when you compare numbers?

▶ What is the highest number your team could roll with two dot cubes?

▶ What is the highest number your team could roll with three dot cubes?

Encourage student discussion of these questions and answers.

Progress Monitoring	
If... students need assistance verbalizing this game,	▶ **Then...** partner them with students who have more advanced English comprehension.

4 ASSESS

Informal Assessment

Have students complete the activity below to make sure they understand the vocabulary. As students use each word:

1. Check understanding.

2. Correct errors.

3. Recheck for understanding.

- Have students define the word *compare* in their own words and describe how they compared numbers during the game.

- Have students roll two dot cubes and determine the sum. Have students define *sum* in terms of the numbers rolled.

For each word, use the following rubric to assign a score.

The student can repeat the word when prompted. (1 point)

The student knows the word but does not know its meaning. (2 points)

The student has a vague idea of the word's meaning. (3 points)

The student knows the word and can use the word in context. (4 points)

Subtracting to Zero

Week 22

Objective
Students can understand the meaning of the term *zero*, can add and subtract whole numbers, and can create word problems.

Vocabulary
- **add** To combine numbers or put together numbers
- **equal** Having the same amount; identical in value or notation
- **equal sign** A symbol that means "having the same amount"
- **equation** A number story
- **minus sign** A symbol that means "take away"
- **subtract** To take away, as in one quantity from another
- **zero** None; the number that, when used as an addend, leave any number unchanged

Materials
Program Materials
- Vocabulary Cards: *add, equal, subtract*
- Two-color counters

1 WARM UP

Introduce each vocabulary word to students. Say the word aloud and have students repeat it.

After defining the vocabulary words, show students the *add, equal,* and *subtract* **Vocabulary Cards** while saying each word aloud.

As a group, review the meaning of *zero*. Invite students to help you make a chart with different words for *zero*, such as *none* and *nothing*. Include translations for the term from students' primary languages.

2 ENGAGE

Tell and demonstrate a series of stories about quantities being taken away. You may use the following stories or make up your own. After each story, write the corresponding equation.

▶ **I had three cookies in my lunch.**

▶ **My sister came along and ate them all.**

▶ **How many cookies do I have now?** 0

▶ **I made five presents for my family.**

▶ **I gave all five presents away.**

▶ **How many presents do I have now?** 0

▶ **One day I looked out my window and saw nine birds sitting on a fence.**

▶ **Suddenly a cat came along, and nine birds flew away.**

▶ **How many birds were still sitting on the fence?** 0

When students are ready, have them create their own subtraction word problems. Organize students into pairs, and distribute one formal subtraction problem to each pair.

Tell students that you want them to make up their own stories about what happened and to draw pictures that show the sequence of actions. Allow students time to make up their stories, and then invite each pair to present its story to the group, providing the answer to the problem as it does so.

Teacher Note

English learners may benefit from additional practice of the /z/ sound in *zero*. Encourage them to think of sounds that use *z*, such as *buzz, zoom, zap, zebra,* and so on.

Progress Monitoring

If... students need help constructing their word problems,	▶ Then... provide them with a key word or subject for their story.

3 REFLECT

Extended Response

▸ **How would you describe this game to someone new in our class?**

▸ **Describe a time outside of class when you had zero of something.**

▸ **Does a number minus zero always equal itself? Why?**

Encourage student discussion of these questions and answers.

Progress Monitoring	
If... students can fluently construct a story to describe the formal equation,	▸ **Then...** challenge them to write their own equations and to make up stories to describe them.

4 ASSESS

Informal Assessment

Have students complete the activity below to make sure they understand the vocabulary. As students use each word:

1. Check understanding.
2. Correct errors.
3. Recheck for understanding.

- Have students define *zero* and describe a number story that includes a zero value.
- Have students model a subtraction problem with a difference of zero.

For each word, use the following rubric to assign a score.

The student can repeat the word when prompted. (1 point)

The student knows the word but does not know its meaning. (2 points)

The student has a vague idea of the word's meaning. (3 points)

The student knows the word and can use the word in context. (4 points)

More Adding and Subtracting

Week 23

Objective
Students can continue to recognize the number sequence and can understand the meanings of the terms *plus sign, minus sign,* and *equal sign.*

Vocabulary
- **equal** Having the same amount; identical in value or notation
- **equal sign** A symbol that means "having the same amount"
- **minus sign** A symbol that means "take away"
- **missing number** The number needed to complete an equation
- **plus sign** A symbol that means "put together"
- **subtract** To take away, as in one quantity from another

Materials
Program Materials
- Vocabulary Card: *add, equal, subtract*
- Number Cards (1–10), p. 129
- Number Cards (11–20), p. 130
- Plus, Minus, Equal, p. 137

1 WARM UP

Introduce each vocabulary word to students. Say the word aloud and have students repeat it.

After defining the vocabulary words, show students the *add, equal,* and *subtract* **Vocabulary Cards** while saying each word aloud.

Organize students in pairs or small groups. Distribute a shuffled pile of Number Cards (1–10) and Number Cards (11–20) to each group of students, and have students arrange the cards in descending order from 20 to 1. Each group should count the numbers aloud. If students need help, encourage them to discuss their question with other students.

2 ENGAGE

Distribute a copy of Plus, Minus, Equal, p. 137, to each student. Write *3 _____ 5 = 8* on the board. Invite students to take turns identifying the numbers and correct math symbol needed to create an equation. Have students say the equation aloud.

▶ **How do you know what symbol to use?**
Possible answer: The answer was larger than the first number, so I knew I needed to add.

▶ **How can you check your answer?** Possible answer: I could subtract 5 from 8 to see if I get 3.

Teacher Note

Repetition is a key element in learning addition and subtraction facts. For English learners, it is also a key to learning the labels and language needed to discuss these concepts. Help students construct a model of an addition fact with a missing addend and a related subtraction fact. Label each element to remind English learners what each element is called. Be sure to include *addend, missing addend, plus sign, equal sign,* and *sum.*

Progress Monitoring

If... you think students are comfortable identifying math symbols for equations,	▶ **Then...** write equations with missing numbers instead of missing symbols, and have students solve the equations.

3 REFLECT

Extended Response

▶ **What symbol do we use to show we are adding something?**

▶ **What symbol do we use to show we are subtracting something?**

▶ **What helps you determine a missing number?**

Encourage student discussion of these questions and answers.

Progress Monitoring	
If... students are puzzled by the variety of words used to describe subtraction, such as *take away*, *minus*, *less*, and *subtracted from*,	▶ **Then...** help them brainstorm a list of words and phrases used to describe subtraction.

4 ASSESS

Informal Assessment

Have students complete the activity below to make sure they understand the vocabulary. As students use each word:

1. Check understanding.

2. Correct errors.

3. Recheck for understanding.

- Have students draw a plus sign, a minus sign, and an equal sign and describe how they are used in equations.

- Have students define *equal* in their own words and describe how they know when one value is equal to another.

For each word, use the following rubric to assign a score.

The student can repeat the word when prompted. (1 point)

The student knows the word but does not know its meaning. (2 points)

The student has a vague idea of the word's meaning. (3 points)

The student knows the word and can use the word in context. (4 points)

Week 24

Objective
Students can identify numerals from 1 to 100 and can compare and order numbers.

Vocabulary
- **missing number** The number needed to complete an equation
- **pattern** Something that repeats the same way each time
- **position** Where something is; the location of something in relationship to other things

Materials
Additional Materials
- 10 index cards
- number line drawn on long sheet(s) of paper taped around the perimeter of the room

1 WARM UP

Introduce each vocabulary word to students. Say the word aloud and have students repeat it.

Label the 10 index cards 1–9, 10–19, 20–29, 30–39, and so on to 90–100. Distribute one index card to each of ten students, and have them count aloud to 100 by referencing their index cards. If the class does not have ten students, let some students hold more than one card.

▶ **What pattern do you notice when you count aloud to 100?**

2 ENGAGE

Create a large number line on paper, and tape it around the perimeter of the room. Label points at appropriate intervals (0, 10, 20, 30, and so on to 100). Point out the pattern of sets of ten. Help students work as a group to label the remaining numbers on the number line.

Call on students to approach the number line individually and answer questions about numerals and their positions.

▶ **Find 55. Use the words *before*, *after*, and *between* to describe the position of 55.**

▶ **What number is before 75? After 75?**

▶ **What number is closest to 100?**

▶ **What is the position of the number 1?**

Teacher Note

Some English learners may recognize the larger numerals on the number line but may need reference for pronunciation. Beneath the appropriate points, write *Twenty, Thirty, Forty*, and so on. Students may not need to use these prompts but could benefit by recognizing the written words.

Progress Monitoring

If... students are confident with identifying numerals on the number line,	▶ Then... write number sequences on the board with missing numbers and ask students to identify the missing number.

3 REFLECT

Extended Response

▶ Describe a time when you have seen 100 of something.

▶ Describe the position of 50 on the number line.

▶ What patterns do you see on the number line?

Encourage student discussion of these questions and answers.

Progress Monitoring	
If... students struggle with the lesson vocabulary,	▶ **Then...** provide some one-on-one practice or help students use sentences to describe numbers on the number line.

4 ASSESS

Informal Assessment

Have students complete the activity below to make sure they understand the vocabulary. As students use each word:

1. Check understanding.

2. Correct errors.

3. Recheck for understanding.

- Have student define *position* and name some position words.
- Point to a number on a number line. Have students describe the number using position words such as *before, after,* and *between.*

For each word, use the following rubric to assign a score.

The student can repeat the word when prompted. (1 point)

The student knows the word but does not know its meaning. (2 points)

The student has a vague idea of the word's meaning. (3 points)

The student knows the word and can use the word in context. (4 points)

More Numbers to 100

Week 25

Objective
Students can understand the meanings of the terms *before*, *after*, *near*, *far*, and additional comparative and superlative terms in relation to the number sequence.

Vocabulary
- **after** Behind; coming later
- **before** In front of; coming first
- **far** Not close to; a long way away
- **near** Close to
- **ones column** The numeral on the far right of a number that tells how many ones are in the number
- **tens column** The numeral to the left of the ones column that tells how many tens are in a number

Materials

Program Materials
Comparison Table, p. 138

Additional Materials
number line with points labeled from 1 to 100 taped to perimeter of classroom

1 WARM UP

Introduce each vocabulary word to students. Say the word aloud and have students repeat it.

Distribute a copy of Comparative Table, p. 138, to each student. Help students complete the worksheet and discuss the terms until students demonstrate understanding.

2 ENGAGE

Engage students in a group conversation about the terms *near, far, nearest, farthest, closer to,* and *farther from.* Ask students to generate sentences using these terms, such as *My desk is near the door but farthest from the window.* As students use the vocabulary, write their sentences on the board.

After everyone has shared at least one sentence using the vocabulary, ask them questions about the number line such as the following.

► **Go to 63. Is 63 before or after 64?**

► **How many tens are in this number?** 6

► **How did you figure that out?** Answers will vary but may indicate knowledge that the 6 is in the tens column or that the number 63 includes 6 groups of ten.

► **Is 63 closer to 0 or 100?** 100

► **Is 63 farther from 50 or 100?** 100

Teacher Note

English learners often understand math concepts better when they can interact with other students who speak the same primary language. Allow children to work with partners or cross-age helpers who speak the same primary language so they can check understanding with one another in team situations.

Progress Monitoring

If... students need help using or explaining their comparative words,

► **Then...** encourage them to draw pictures and/or translations on their Comparison Table to help them use the vocabulary in mathematical sentences.

3 REFLECT

Extended Response

► **What number is in the ones column in 26?**

► **What number is in the tens column in 45?**

► **What are some things that are found near your home?**

► **What are some numbers near 27?**

Encourage student discussion of these questions and answers.

Progress Monitoring

If... students struggle with the new vocabulary,	► **Then...** offer them a sentence frame such as *I am sitting _____ the door* or *The playground is _____ the gym.*

4 ASSESS

Informal Assessment

Have students complete the activity below to make sure they understand the vocabulary. As students use each word:

1. Check understanding.

2. Correct errors.

3. Recheck for understanding.

- Write a two-digit number on the board. Have students identify the digit in the ones column and the digit in the tens column.

- Point to a number on the number line. Have students describe other numbers using the terms *before, after, near,* and *far* in relationship to the given number.

For each word, use the following rubric to assign a score.

The student can repeat the word when prompted. (1 point)

The student knows the word but does not know its meaning. (2 points)

The student has a vague idea of the word's meaning. (3 points)

The student knows the word and can use the word in context. (4 points)

Final Oral Assessment

Administer the appropriate Final Oral Assessment, pp. 88–89, to each student. Use the rubric to determine students' levels of vocabulary acquisition.

Use the Student Assessment Record, page 142, to record the assessment results.

Final Oral Assessment 1, p. 88

Weeks 21–25 Final Oral Assessment 1 (Beginning English Learners)

Directions: Read each question to the student, and record his or her oral responses.
Some questions have teacher directions. Teacher directions are indicated in italics.
Allow students to use pencil and paper to work their responses.

1. **Compare** these numbers. Which one is bigger? *Write 24 and 42 on a piece of paper.* **42 is bigger**

2. **Compare** these numbers. Which number is highest? *Write 16, 23, and 44 on a piece of paper.* **44**

3. What does *zero* mean? **none or nothing**

4. **Subtract.** What is the answer? *Write 16 − 16 = _____ on a piece of paper.* **0**

5. Is this a **plus sign** or a **minus sign?** *Write a plus sign on a piece of paper.* **plus sign**

6. Is this a **plus sign** or a **minus sign?** *Write a minus sign on a piece of paper.* **minus sign**

7. What does a **minus sign** mean? **to subtract or take away**

8. What math symbol will you use to write this **equation?** *Write 4 _____ 5 = 9 on a piece of paper.* **plus sign**

9. Solve this **equation.** *Write 5 − 2 = _____ on a piece of paper.* **3**

10. Solve this **equation.** *Write 14 − 7 = _____ on a piece of paper.* **7**

- **Minimal Understanding:** 0–3 of Questions 1–10 correct
- **Basic Understanding:** 4–7 of Questions 1–10 correct
- **Secure Understanding:** 8–10 of Questions 1–10 correct

Use the Student Assessment Record, page 142, to record the assessment results.

Weeks 21–25 Final Oral Assessment 2 (Intermediate and Advanced English Learners)

Directions: Read each question to the student, and record his or her oral responses. Some questions have teacher directions. Teacher directions are indicated in italics. Allow students to use pencil and paper to work their responses.

1. What math symbol would you use to write this **equation**? *Write 8 _____ 6 = 2 on a piece of paper.* **—**

2. What numbers come **after** 40? **41, 42, 43, 44, 45, and so on**

3. What digit is in the **tens column** in this number? *Write 100 on a piece of paper.* **0**

4. Is 40 **near** 50 or 100? **50**

5. **Between** what numbers is 50? **49 and 51**

6. What is the number **between** 64 and 66? *Write 62, 63, 64, _____, 66 on a piece of paper.* **65**

7. What are the numbers **between** 45 and 48? *Write 44, 45,_____ ,_____ , 48 on a piece of paper.* **46, 47**

8. **Add.** Find the missing number. *Write 4 + _____ = 10 on a piece of paper.* **6**

9. Solve this **equation.** *Write 13 − 6 = _____ on a piece of paper.* **7**

10. Solve this **equation.** *Write 12 − 3 = _____ on a piece of paper.* **9**

- **Minimal Understanding:** 0–3 of Questions 1–10 correct
- **Basic Understanding:** 4–7 of Questions 1–10 correct
- **Secure Understanding:** 8–10 of Questions 1–10 correct

Use the Student Assessment Record, page 142, to record the assessment results.

Introduction

Weeks 26–30

Section at a Glance

In this section, students will learn the vocabulary associated with **Number Worlds,** Level C, Weeks 26–30. Students are expected to understand and use math symbols, understand the meaning of *equality*, add and subtract whole numbers, solve word problems, count and identify numbers to 100, skip count by tens, write addition and subtraction equations, use spatial terms to describe a location, and create a map. Before beginning the section, assess students' general knowledge of math vocabulary using the Individual Oral Assessment on page 91.

How Students Learn Vocabulary

One strategy for helping English learners comprehend new math concepts is using models and demonstrations. Number lines provide English learners with physical evidence of a number sequence when adding and subtracting whole numbers.

Academic Vocabulary Taught in Weeks 26–30

Week 26

add To combine numbers or put together numbers

altogether In all; total

equal Having the same amount; identical in value or notation

equal sign A symbol that means "having the same amount"

equation A number story

plus sign A symbol that means "add"

sum The answer to an addition problem

word problem A problem that has a story as well as numbers

Week 27

add To combine numbers or put together numbers

ones Single units

ones column The numeral on the far right of a number that tells how many ones are in the number

subtract To take away, as in one quantity from another

tens column The numeral to the left of the ones column that tells how many tens are in a number

tens Groups of ten single units

Week 28

add To combine numbers or put together numbers

compare To think about how things are the same and different

measure To find out how long or how tall something is

minus sign A symbol that means "take away"

subtract To take away

solve To figure out; to find the answer

sum The answer to an addition problem

Week 29

altogether In all; total

equation A number story

left over Not used

word problem A problem that has a story as well as numbers

Week 30

after Behind; coming later

before In front of; coming first

direction A way you move, such as right or left, up or down

far Not close; a long way

Weeks 26–30 Individual Oral Assessment

Directions: Read each question to the student, and record his or her oral responses. Some questions have teacher directions. Teacher directions are indicated in italics. Allow students to use pencil and paper to work their responses.

1. Does a **plus sign** mean "add"? **yes**

2. Does a **minus sign** mean "equal"? **no**

3. Does *sum* mean "total"? **yes**

4. Is this a **word problem?** *Write a word problem on the board.* **yes**

5. Is this a **word problem?** *Write 7 + 3 = 10 on the board.* **no**

6. What number is **before** 20? *Draw a number line with points labeled from 1 to 20.* **19**

7. Which number is **far** from 0? *Write 2 and 34 on a piece of paper.* **34**

8. What number is in the **tens column?** *Write 86 on a piece of paper.* **8**

9. What number is in the **ones column?** *Write 63 on a piece of paper.* **3**

10. If I have two dogs and my friend has three dogs, how many dogs do we have **altogether?** **2 + 3 = 5 dogs**

11. Find the **sum.** *Write 12 + 7 = _____ on a piece of paper.* **12 + 7 = 19**

12. Find the **sum.** *Write 8 + 5 = _____ on a piece of paper.* **8 + 5 = 13**

13. **Solve** this **equation.** You may draw a picture. *Write 12 − 7 = _____ on a piece of paper.* **12 − 7 = 5**

- **Beginning English Learners:** 0–3 of Questions 1–10 correct
- **Intermediate English Learners:** 4–7 of Questions 1–10 correct
- **Advanced English Learners:** 8–10 of Questions 1–10 correct
- If the student is able to answer Questions 11–13, then he or she can understand the mathematics taught in this unit but may still have difficulty with the academic vocabulary.

Use the Student Assessment Record, page 142, to record the assessment results.

Addition Stories

Week 26

Objective
Students can add and subtract whole numbers by writing and drawing word problems.

Vocabulary
- **add** To combine numbers or put together numbers
- **altogether** In all; total
- **equal** Having the same amount; identical in value or notation
- **equal sign** A symbol that means "having the same amount"
- **equation** A number story
- **plus sign** A symbol that means "add"
- **sum** The answer to an addition problem sum
- **word problem** A problem that has a story as well as numbers

Materials
Program Materials
Vocabulary Cards: *add, equal, sum*

Additional Materials
index cards with addition problems written on them (optional)

1 WARM UP

Introduce each vocabulary word to students. Say the word aloud and have students repeat it.

Hold up the *add, equal,* and *sum* **Vocabulary Cards.** Read each word aloud, and have students repeat the words.

2 ENGAGE

Write and draw word problems, one per sheet of paper, and post the word problems around the room. Organize students into pairs or small groups to create equations and solve each word problem. Encourage students to draw the word problem on their own papers and discuss it with their groups.

Examples of some word problems include the following:

▶ **I invited 5 people to my party. Then I invited 3 more. If I want each of my guests to have a prize, how many prizes will I need? Write an equation and find the sum.** $5 + 3 = 8$ prizes

▶ **My house is 1 block from the store. The store is 8 blocks from my grandmother's house. If I go to the store and then to my grandmother's house, how many blocks will I walk altogether?** $1 + 8 = 9$ blocks

▶ **I bought a fish tank and 14 fish. My mother gave me 4 more fish. How many fish do I have altogether?** $14 + 4 = 18$ fish

▶ **I picked 8 apples from one tree and 3 apples from another tree. How many apples did I pick altogether?** $8 + 3 = 11$ apples

▶ **Ask students to share their equations with the class.**

▶ **How did you write your equation?**

▶ **What math signs did you use?**

Teacher Note

Students may want to create their own equations once they are comfortable with this group game. Allow them to work in pairs to create a word problem based on an addition equation.

Progress Monitoring

If... students are having difficulty creating their own number stories, ▶ **Then...** distribute addition problems written on index cards to pairs of students. Have students create a word story using the problem on the card as a guide.

3 REFLECT

Extended Response

▶ What do you like about word problems?

▶ What is difficult about solving word problems?

▶ How do you know what math signs to use to find a sum in word problems?

Encourage student discussion of these questions and answers.

Progress Monitoring	
If... students are having difficulty creating equations from word problems,	▶ **Then...** create simple word problems and discuss possible corresponding equations with students.

4 ASSESS

Informal Assessment

Have students complete the activity below to make sure they understand the vocabulary. As students use each word:

1. Check understanding.

2. Correct errors.

3. Recheck for understanding.

- Have students define *word problem* in their own words and give an example of a word problem.

- Write a word problem on the board. Have students write an appropriate equation to solve the problem.

For each word, use the following rubric to assign a score.

The student can repeat the word when prompted. (1 point)

The student knows the word but does not know its meaning. (2 points)

The student has a vague idea of the word's meaning. (3 points)

The student knows the word and can use the word in context. (4 points)

Tens and Ones

Week 27

Objective
Students can count and identify numbers to 100, skip count by tens, and understand the meanings of the terms *ones column* and *tens column*.

Vocabulary
- **add** To combine numbers or put together numbers
- **ones** Single units
- **ones column** The numeral on the far right of a number that tells how many ones are in the number
- **subtract** To take away, as in one quantity from another
- **tens column** The numeral to the left of the ones column that tells how many tens are in a number
- **tens** Groups of ten single units

Materials

Program Materials
Vocabulary Cards: *add, ones, subtract, tens*

Additional Materials
- paper number line labeled from 1 to 100 taped to perimeter of classroom
- index cards
- opaque bag or shoe box

1 | WARM UP

Introduce each vocabulary word to students. Say the word aloud and have students repeat it.

After defining the vocabulary words, show students the *add, ones, subtract,* and *tens* **Vocabulary Cards** while saying each word aloud.

Write various numbers on the board, and discuss with students the digits in the ones column and the tens column. Ask students how many ones and tens are in the given numbers.

2 | ENGAGE

Create a paper number line with points labeled from 0 to 100, and tape it to the perimeter of your classroom. Label index cards 0, 10, 20, 30, 40, and so on to 100. Display these cards as you have the students skip count by tens. Place these cards on the number line, and then have the students skip count again.

On index cards, write random numbers between 1 and 100, and ensure that you have at least as many cards as students. Place the cards in a bag, and ask a student to draw an index card from the bag. Ask the following questions of each student.

- ▶ **What number do you have?**
- ▶ **Find and label that number on the number line.**
- ▶ **How many tens are in your number?**
- ▶ **How many ones are in your number?**
- ▶ **Start at 0, and skip count by tens toward your number. When you have counted all the tens, count on by ones until you reach your number.**

Continue this game until everyone has had a chance to find and identify a number on the number line.

Teacher Note

For an added challenge to the game, have a second student find his or her number from the location of the first student's number. Ask questions such as *What direction will you move to find your number?* and *How did you find your number?*

Progress Monitoring

If... students are confident with this game,	▶ Then... organize students into small groups and allow them to play the game by adding and subtracting numbers on the number line.

3 REFLECT

Extended Response

▶ **Where is the ones column in a number?**

▶ **What does *ones column* mean?**

▶ **Where is the tens column in a number?**

▶ **What does *tens column* mean?**

Encourage student discussion of these questions and answers.

Progress Monitoring	
If... English learners are uncomfortable standing in front of the class to identify numbers on the number line,	▶ **Then...** have students work in small groups. Visit each group to ask questions and assess understanding.

4 ASSESS

Informal Assessment

Have students complete the activity below to make sure they understand the vocabulary. As students use each word:

1. Check understanding.
2. Correct errors.
3. Recheck for understanding.

- Write a two-digit number on the board. Have students identify the digit in the ones column and the digit in the tens column.

- Have students describe how they can use their knowledge of the ones column and the tens column to locate a number on the number line.

For each word, use the following rubric to assign a score.

The student can repeat the word when prompted. (1 point)

The student knows the word but does not know its meaning. (2 points)

The student has a vague idea of the word's meaning. (3 points)

The student knows the word and can use the word in context. (4 points)

Adding and Subtracting Length

Week 28

Objective
Students can count aloud to 100 and can use subtraction to determine how much longer one item is than another.

Vocabulary
- **add** To combine numbers or put together numbers
- **compare** To think about how things are the same and different
- **measure** To find out how long or how tall something is
- **minus sign** A symbol that means "take away"
- **subtract** To take away
- **solve** To figure out; to find the answer
- **sum** The answer to an addition problem

Materials

Program Materials	Additional Materials
• Vocabulary Card: *add, subtract*	small bean bag or soft ball to toss
• Measuring Length, p. 139	
• Learning Links	

1 WARM UP

Introduce each vocabulary word to students. Say the word aloud and have students repeat it.

After defining the vocabulary words, show students the *add* and *subtract* **Vocabulary Cards** while saying each word aloud.

Explain to students that they will take turns passing the bean bag to each other, counting aloud by ones such as 1–10, pass, 11–20, pass, 21–30, and so on. For variation, have students skip count aloud by tens. If a student makes a counting error, have him or her re-count that segment of the sequence and ask other students for help.

2 ENGAGE

Distribute a copy of Measuring Length, p. 139, and five to ten Learning Links to each student. Organize students into small groups. Have students use Learning Links to measure certain objects around the room and record the name of the object and its length on their worksheets.

Some objects that students might measure include a desk, the height of the classroom doorknob, a student chair, the length of each partner's shoe, and so on. Label these objects so students can copy the object names onto their worksheets.

Ask pairs to compare their measurements with the class.

► **How much taller is (the doorknob) than (the desk)? Use subtraction to solve this problem.**

► **What is the sum of the heights of (the chair) and (the desk)?**

► **How much longer is (the first partner's shoe) than (the second partner's shoe)?**

► **How did you figure out your answers?**

Teacher Note

This activity can be challenging for English learners, as it requires translation of object names as well as equation writing. Encourage students to draw a picture of each object they measure and include the drawings in each equation if it helps them determine the answer.

Progress Monitoring

If... think students are struggling with this activity,	► **Then...** label each object with its measurement, and ask students to focus on creating the equation instead of measuring the objects.

3 REFLECT

Extended Response

▶ What does *longer* mean?

▶ How can we find out how much longer an object is than another object?

▶ What symbols will you use to solve this problem?

▶ Describe a time when you have measured something.

Encourage student discussion of these questions and answers.

Progress Monitoring

| **If...** students have trouble writing subtraction equations, | ▶ **Then...** remind them that they will always subtract the smaller number from the bigger number to find the difference in length or height. |

4 ASSESS

Informal Assessment

Have students complete the activity below to make sure they understand the vocabulary. As students use each word:

1. Check understanding.

2. Correct errors.

3. Recheck for understanding.

- Have students define measure in their own words and describe how they *measured* the objects in the lesson.

- Have students describe how they wrote and solved subtraction problems related to the objects they measured.

For each word, use the following rubric to assign a score.

The student can repeat the word when prompted. (1 point)

The student knows the word but does not know its meaning. (2 points)

The student has a vague idea of the word's meaning. (3 points)

The student knows the word and can use the word in context. (4 points)

Addition and Subtraction Stories

Week 29

Objective
Students can reinforce counting and number sequence skills and solve word problems.

Vocabulary
- **altogether** In all; total
- **equation** A number story
- **left over** Not used
- **word problem** A problem that has a story as well as numbers

Materials
paper and pencils for drawing

1 WARM UP

Introduce each vocabulary word to students. Say the word aloud and have students repeat it.

As a group, play "What Number Am I?" Present students with mystery problems. Ask students to use number lines to help solve the mysteries. Sample problems include the following:

- ▶ **I come two numbers after 5. I come one number before 8. I am one more than 6. What number am I?** 7

- ▶ **I am more than 10. I am less than 15. The number right before me is 11. The number right after me is 13. What number am I?** 12

- ▶ **I am less than 20. I am more than 18. What number am I?** 19

- ▶ **I am one group of 10 and 4 ones. I am more than 12, and I am more than 13. I am less than 16, and I am less than 15. What number am I?** 14

2 ENGAGE

Write the elements of a word problem on the board: *3 birds, 2 birds, 5 birds.*

As a group, construct a word problem with these subjects. Your word problem might resemble the following example.

- ▶ **I looked in the tree and saw 3 birds. Two more birds joined them. There were 5 birds altogether.**

Write the elements of a new word problem on the board: *6 hamburgers, 4 hamburgers, 2 hamburgers.*

- ▶ **Mom made 6 hamburgers, and we ate 4. How many are left over?**

Invite students to work in pairs to craft word problems with the elements you provide. Create word problem foundations for each group, and encourage them to draw pictures of their equations for visual reference.

Teacher Note

Model the problem-solving process as often as needed to help student understanding.

Progress Monitoring

If... students are challenged by this activity,	▶ **Then...** remind them to create an equation from the word problem objects first and decide what kind of language they will use to make an addition or a subtraction word problem.

3 REFLECT

Extended Response

▶ What words help you make an addition word problem?

▶ What words help you make a subtraction word problem?

▶ What helps you create word problems?

▶ What helps you solve word problems?

Encourage student discussion of these questions and answers.

Progress Monitoring	
If... students need an additional challenge,	▶ **Then...** ask them to create their own word problems without prompting.

4 ASSESS

Informal Assessment

Have students complete the activity below to make sure they understand the vocabulary. As students use each word:

1. Check understanding.

2. Correct errors.

3. Recheck for understanding.

- Have students define *word problem* and give an example of a word problem.

- Have students define *altogether* and explain how they would use the term to describe an addition word problem.

For each word, use the following rubric to assign a score.

The student can repeat the word when prompted. (1 point)

The student knows the word but does not know its meaning. (2 points)

The student has a vague idea of the word's meaning. (3 points)

The student knows the word and can use the word in context. (4 points)

Making a Map

Week 30

Objective
Students can compare and order numbers, use spatial terms to describe location, and create a map.

Vocabulary
- **after** Behind; coming later
- **before** In front of; coming first
- **direction** A way you move, such as right or left, up or down
- **far** Not close; a long way

Materials

Program Materials
Map It, p. 140

Additional Materials
small bean bag or soft ball to toss

1 WARM UP

Introduce each vocabulary word to students. Say the word aloud and have students repeat it.

- ▶ **What is a map?** a drawing of an area
- ▶ **What does it show us?** Possible answers: streets, a city
- ▶ **Can you use a map to see how far it is between places?** yes
- ▶ **Can you use a map to follow directions?** Yes, you can see the places as you go right or left, up or down.
- ▶ **When have you used a map before?** Answers will vary.

2 ENGAGE

Distribute a copy of Map It, p. 140, to each student. Label the desks in order from 1 to 5.

- ▶ **How many desks will you pass on the way to Desk 5?** four desks
- ▶ **What desk is before Desk 4?** Desk 3
- ▶ **What desk is between Desk 1 and Desk 3?** Desk 2
- ▶ **What desk is after Desk 3?** Desk 4
- ▶ **How far is it from Desk 2 to Desk 5?** 3 desks
- ▶ **How can I get from Desk 4 to Desk 1?** Go back 3

You can repeat this activity by labeling the desks in different ways, such as numerically (1, 2, 3, 4, 5), with student's names (James, Tia, Paul, Juan, Becky), and color of the desks (red, yellow, purple, orange, green). Ask similar questions about direction on the map.

When students have completed this activity comfortably, encourage them to draw their own maps of the classroom, hallway, or other school setting. Have students give you directions from one place to another. Ask questions about position, direction, far, and near.

Teacher Note

Students may want to work in pairs on this activity. One student will describe directions to a desk, and the other student will follow those directions and ask questions.

Progress Monitoring

If... students can create an accurate hallway map with directions,	▶ Then... challenge them to map other things, such as the route they take to school or the areas of the classroom. Have students write directions to one of the areas on the map (without identifying the area) and then trade maps with a classmate to see if that student can use the directions to locate the correct area.

3 REFLECT

Extended Response

▶ Describe a time when you would use a map.

▶ How would you give someone directions to walk from the gym to this classroom? Is it before or after the music room? Is it far from the cafeteria? What direction should I go first?

Encourage student discussion of these questions and answers.

Progress Monitoring	
If... you realize that a student has misunderstood a new vocabulary word,	▶ **Then...** help the student with some additional one-on-one practice. Using it in free discussion is productive only when the student understands its meaning.

4 ASSESS

Informal Assessment

Have students complete the activity below to make sure they understand the vocabulary. As students use each word:

1. Check understanding.

2. Correct errors.

3. Recheck for understanding.

- Have students define *direction* in their own words and give examples of ways to give direction.

- Have students describe locations on a map using the terms *before, after, direction,* and *far.*

For each word, use the following rubric to assign a score.

The student can repeat the word when prompted. (1 point)

The student knows the word but does not know its meaning. (2 points)

The student has a vague idea of the word's meaning. (3 points)

The student knows the word and can use the word in context. (4 points)

Final Oral Assessment

Administer the appropriate Final Oral Assessment, pp. 102–103, to each student. Use the rubric to determine students' levels of vocabulary acquisition.

Use the Student Assessment Record, page 142, to record the assessment results.

Final Oral Assessment 1, p. 102

Weeks 26–30 Final Oral Assessment 1 (Beginning English Learners)

Directions: Read each question to the student, and record his or her oral responses. Some questions have teacher directions. Teacher directions are indicated in italics. Allow students to use pencil and paper to work their responses.

1. How much taller is the big dog than the small dog? **Subtract.** *Draw a big dog, and draw six Learning Links beside it. Draw a small dog, and draw three Learning Links beside it.* **3**

2. Write the **equation** that shows how much taller the big dog is. **6 − 3 = 3**

3. Solve this **equation.** *Write* 14 + 4 = _____ *on a piece of paper.* **14 + 4 = 18**

4. Solve this **equation.** *Write* 14 − 4 = _____ *on a piece of paper.* **14 − 4 = 10**

5. Solve this **word problem.** I had seventeen newspapers and delivered all of them. How many newspapers do I have left? **0**

6. Write an **equation** for the previous word problem. **17 − 17 = 0**

7. What number is in the **ones column?** *Write* 38 *on the board.* **8**

8. What number is in the **tens column?** *Write* 41 *on the board.* **4**

9. Write an **equation** to find out how much taller the chair is than the table. *Draw a chair, and label it 8. Draw a table, and label it 5.* **8 − 5 = 3**

10. Solve this **equation.** *Write* 12 + 8 = _____ *on a piece of paper.* **12 + 8 = 20**

- **Minimal Understanding:** 0–3 of Questions 1–10 correct
- **Basic Understanding:** 4–7 of Questions 1–10 correct
- **Secure Understanding:** 8–10 of Questions 1–10 correct

Use the Student Assessment Record, page 142, to record the assessment results.

Weeks 26–30 Final Oral Assessment 2 (Intermediate and Advanced English Learners)

Directions: Read each question to the student, and record his or her oral responses. Some questions have teacher directions. Teacher directions are indicated in italics. Allow students to use pencil and paper to work their responses.

1. If I have three balloons and you have four balloons, how many balloons do we have **altogether?** *seven balloons*

2. How would you write an **equation** for the previous word problem? **3 + 4 = 7**

3. I had eight model cars. I gave two to my friend. How many model cars are **left over?** Write the equation. **8 − 2 = 6**

4. I had twelve flowers. I gave nine to my brother. How many flowers are **left over?** Write the **equation.** **12 − 9 = 3**

5. Joe has four cats. Cara has two cats. How many cats do they have **altogether?** Write the **equation.** **4 + 2 = 6**

6. Francis has eight books. Carlos has four books. How many books do they have **altogether?** Write the equation. **8 + 4 = 12**

7. What door is **before** Door 3? *Draw 4 doors on a piece of paper. Label them 1, 2, 3, 4. Draw a "Start" box to the far left of Door 1.* **Door 2**

8. What door is **after** Door 3? **Door 4**

9. Solve this **equation.** *Write* 3 + 5 + 9 = _____ *on a piece of paper.* **3 + 5 + 9 = 17**

10. Solve this **equation.** *Write* 6 − 2 − 1 = ____ *on a piece of paper.* **6 − 2 − 1 = 3**

- **Minimal Understanding:** 0–3 of Questions 1–10 correct
- **Basic Understanding:** 4–7 of Questions 1–10 correct
- **Secure Understanding:** 8–10 of Questions 1–10 correct

Use the Student Assessment Record, page 142, to record the assessment results.

Weeks 31–32

Section at a Glance

In this section, students will learn the vocabulary associated with **Number Worlds,** Level C, Weeks 31–32. Students will learn about the analog clock and the terms *clock, hour,* and *half hour* as commonly used periods of time. They will learn how to manipulate the hands on a clock in order to show hourly time. Students will also use the words *o'clock, hour hand,* and *minute hand*. Students will also learn about six denominations of money, the penny, the nickel, dime, the $1 bill, the $5 bill, and the $10 bill. They will also role-play using currency and getting change back. Before beginning the section, assess students' general knowledge of math vocabulary using the Individual Oral Assessment on page 105.

How Students Learn Vocabulary

This section's vocabulary addresses time and money. English learners will become familiar with different terms related to time and money using manipulatives such as an analog clock with movable hands and play money. They will also model the concept of subtractions using play money.

Academic Vocabulary Taught in Weeks 31–32

Week 31

clock A special dial for telling the time

hour The time that passes when the hour hand moves from one number to the next number

before At an earlier time

after Following in time

half hour The middle point of an hour

Week 32

$1 bill Paper money worth 100 pennies

$5 bill Paper money worth 5 dollars

$10 bill Paper money worth 10 dollars

penny A small copper coin worth 1 cent

nickel A small silver-colored coin, worth 5 cents

dime The smallest of all coins, silver-colored and worth 10 cents

Weeks 31–32 Individual Oral Assessment

Directions: Read each question to the student, and record his or her oral responses. Some questions have teacher directions. Teacher directions are indicated in italics. Allow students to use pencil and paper to work their responses.

1. Is this a clock? *Show students an analog clock.* **yes**

2. Point to the hour hand. *Show students an analog clock.* **Student should point to the short hand.**

3. Point to the minute hand. **Student should point to the long hand.**

4. What number is the hour hand pointing to? *Move the hour hand so it is pointing to 7. Move the minute hand so it is pointing to 12.* **seven**

5. Show one hour after 2 o'clock. *Set the time so the clock shows 2 o'clock.* **Student should move the hour hand so the clock reads 3 o'clock.**

6. Is this a half hour? *Move the minute hand from the 12 to the 6.* **yes**

7. Is this a nickel? *Place a nickel on the table in front of the student.* **yes**

8. Is this a penny? *Place a dime on the table in front of the student.* **no**

9. Is this a $5 bill? *Place a $1 bill on the table in front of the student.* **no**

10. How many pennies equal one nickel? **five**

11. Subtract two pennies from this set. *Display a pile of five pennies. Give the student a few pennies.* **Student should remove two pennies.**

12. Add one $5 bill to this set. *Display several pennies, nickels, dimes, and a $1 bill. Give the student a few pennies, nickels, dimes, two $1 bills, and a $5 bill.* **Student should add one $5 bill.**

13. What time is it? *Show students an analog clock set to 9 o'clock.* **nine o'clock**

- **Beginning English Learners:** 0–3 of Questions 1–10 correct
- **Intermediate English Learners:** 4–7 of Questions 1–10 correct
- **Advanced English Learners:** 8–10 of Questions 1–10 correct
- If the student is able to answer Questions 11–13, then he or she can understand the mathematics taught in this unit but may still have difficulty with the academic vocabulary.

Use the Student Assessment Record, page 142, to record the assessment results.

Understanding the Analog Clock

Week 31

Objective

Students become acquainted with an analog clock, read times to the hour and half hour on the clock, and learn about the concept of before and after.

Vocabulary

- **clock** A special dial for telling the time
- **hour** The time that passes when the hour hand moves from one number to the next number
- **before** At an earlier time
- **after** Following in time
- **half hour** The middle point of an hour

Materials

Program Materials

- Vocabulary Cards: *after, before, clock, half hour, hour*
- Make-Your-Own Analog Clock, p. 141

Additional Materials

- analog clocks with moveable hands, one for each student
- brad paper fasteners, one for each student
- photos of various items, including clocks

1 WARM UP

Introduce each vocabulary word to students. Say the word aloud and have students repeat it.

Point to the clock in the classroom.

▶ **What is this? clock**

Say *clock,* and have students repeat.

Show several photos of various items, including several clocks. Point to photos one at a time.

▶ **Is this a clock?**

Distribute an analog clock with movable hands to each student. Discuss the attributes of the clocks.

▶ **What shape is the clock?**

▶ **What numbers do you see?**

▶ **Put your finger on 12. Then touch each number and say its name.**

▶ **After 12, what number do you say next?**

2 ENGAGE

Point to the hands on your clock and have students touch the hands on their clocks. Tell students that the short hand is the *hour hand*. Model the word and have students say it. Repeat for *minute hand*.

Have students move the hands so that they are both pointing to 12. Say, *This is 12 o'clock.* Have students repeat. Prompt students to move the hour hand to the next number and say the hour.

▶ **Put the hour hand on the 1. What time is it now? one o'clock**

▶ **How many hours have passed? one**

Tell students that when the hour hand travels from a number to the next number, one hour has passed. Repeat moving the hands and saying the time for each number on the clock and telling how many hours have passed.

Hide your clock from students. Set your clock to 5 o'clock and then show students.

▶ **What time is it now?**

Repeat several more times. Then demonstrate *before* and *after* for several times on your clock. Say, *One hour before 5 o'clock is 4 o'clock.*

Dictate some times and have students set their clocks to that time.

▶ **Show me 6 o'clock.**

▶ **What is two hours before 6 o'clock?**

▶ **What is one more hour after 6 o'clock?**

Use your clock to demonstrate moving the minute hand forward one minute.

▶ **Can you move your minute hand forward by one minute? By ten minutes? By 30 minutes?**

Demonstrate on your clock how to move the minute hand from the 12 to the 6 in a clockwise motion.

▶ **Did I go all the way around or halfway around? halfway around**

Tell students that when the minute hand goes halfway around the clock, a half hour has passed. Model the term and have students repeat. Have students show a half hour on their clocks.

Teacher Note ✐

Have students make their own analog clocks to take home and practice with. Give each student a copy of the Make-Your-Own Clock, p. 141 and one brad paper fastener. Monitor as students write the numbers on their clocks. Demonstrate how to secure the hands onto the clock with a brad paper fastener.

Progress Monitoring

If... students struggle with the concept of *before* and *after*,	▶ **Then...** illustrate the terms by creating a simple sequence of events (e.g., putting on socks and shoes) and asking questions such as *What happens first?* Once students understand the correct order of steps, they will pick up the concept of *before* and *after*.

3 REFLECT

Extended Response

▶ **Is time important? Why?**

▶ **Is a minute a long time or a short time? Why?**

▶ **What takes a minute to do? What takes a half hour? An hour?**

▶ **Give an example of *before* and *after*.**

Encourage student discussion of these questions and answers.

Progress Monitoring

If... students are hesitant to contribute to the discussion,	▶ **Then...** offer choices for each question. For example, *Which takes one minute: eating dinner, putting on your shoes, or watching a movie?*

4 ASSESS

Informal Assessment

Have students complete the activity below to make sure they understand the vocabulary. As students use each word:

1. Check understanding.
2. Correct errors.
3. Recheck for understanding.

- Have students identify an hour, a half hour, and a minute.
- Have students move the hands on the clock to show *before* and *after*.

For each word, use the following rubric to assign a score.

The student can repeat the word when prompted. (1 point)

The student knows the word but does not know its meaning. (2 points)

The student has a vague idea of the word's meaning. (3 points)

The student knows the word and can use the word in context. (4 points)

The Ten-Dollar Bill and the Dime

Week 32

Objective
Students will recognize, examine, and compare the values of the penny, nickel, dime, $1 bill, $5 bill, and $10 bill.

Vocabulary
- **$1 bill** Paper money worth 100 pennies
- **$5 bill** Paper money worth 5 dollars
- **$10 bill** Paper money worth 10 dollars
- **penny** A small copper coin worth 1 cent
- **nickel** A small silver-colored coin, worth 5 cents
- **dime** The smallest of all coins, silver-colored and worth 10 cents

Materials

Program Materials
- Vocabulary Cards: *$1 bill, $5 bill, $10 bill, dime, nickel, penny*
- Grocery Store, p. 135
- Plus, Minus, Equal, p. 137

Additional Materials
- pennies
- nickels
- dimes
- $1 bills
- $5 bills
- $10 bills
- photo of the check-out line at a grocery store, including cashier and customer

1 WARM UP

Introduce each vocabulary word to students. Say the word aloud and have students repeat it.

Give each student a penny, nickel, dime, $1 bill, $5 bill, and $10 bill. Have partners examine each piece and talk about its size and color, and also about any pictures, numbers, or words on each one.

▶ **Which ones are coins? Which ones are paper? Which ones are silver? Which ones are green? Which ones have the word *one*? Which ones have the word *five*? Which ones have the word *ten*?**

Review the value of each coin and bill, including that a nickel equals five pennies or a $10 bill equals two $5 bills. Then have students order the currency from least value to greatest value.

2 ENGAGE

Give each student a copy of Grocery Store, p. 135, and a handful of pennies, nickels, and dimes. Point to the bananas.

▶ **How much are the bananas? three cents**

Have students count out three cents. Point to the pears.

▶ **How much are the pears? three cents**

Have students count out three more cents.

▶ **What is the total amount? six cents**

Demonstrate counting three cents plus three cents. Use pennies and the plus sign and equal sign from Plus, Minus, Equal to show 3 pennies + 3 pennies = 6 pennies. Then show students how to convert five of the pennies into a nickel: 3 pennies + 3 pennies = 1 nickel and 1 penny. Repeat with other examples from the Grocery Store page.

Show a photo of the check-out line at a grocery store. Point out the *cashier* and *customer*. Say each word and have students repeat. Tell students you are going to be a customer. You want to buy some eggs and some mushrooms. Have another student act as the cashier and tell you the total amount; encourage the cashier to show the total using the plus sign and equal sign. Give the cashier some money. The cashier should give you the proper change.

Guide students in a role-play about buying things in a store. Make sure they use vocabulary such as *change back, penny, total,* and so on. Have all students practice simultaneously as you monitor. Then have some volunteers present their role-play for the whole group.

Teacher Note 📝

Offer communication guides for students to use during the role-play. For example:

Customer: I want to buy _____ and _____.

Cashier: That's _____ cents plus _____ cents. The total is _____.

Customer: (Customer gives money to cashier and names it; for example:) Here is a dime.

Cashier: _____ cents minus _____ cents equals _____ cents. You get _____ cents change back.

3 REFLECT

Extended Response

▶ **Would you rather have 100 pennies or 5 dimes? Why?**

▶ **Do you save money? Why?**

▶ **What can you buy with a dollar?**

▶ **Which is greater, one $10 bill or three $5 bills? How do you know?**

Encourage student discussion of these questions and answers.

4 ASSESS

Informal Assessment

Have students complete the activity below to make sure they understand the vocabulary. As students use each word:

1. Check understanding.

2. Correct errors.

3. Recheck for understanding.

- Have students point to the penny, nickel, dime, $1 bill, $5 bill, and $10 bill.

- Have students show which item has the least value and the greatest value.

For each word, use the following rubric to assign a score.

The student can repeat the word when prompted. (1 point)

The student knows the word but does not know its meaning. (2 points)

The student has a vague idea of the word's meaning. (3 points)

The student knows the word and can use the word in context. (4 points)

Final Oral Assessment

Administer the appropriate Final Oral Assessment, pp. 110–111, to each student. Use the rubric to determine students' levels of vocabulary acquisition.

Use the Student Assessment Record, page 142, to record the assessment results.

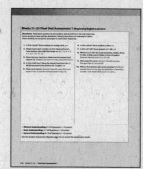

Final Oral Assessment 1, p. 110

Weeks 31–32 Final Oral Assessment 1 (Beginning English Learners)

Directions: Read each question to the student, and record his or her oral responses. Some questions have teacher directions. Teacher directions are indicated in italics. Allow students to use pencil and paper to work their responses.

1. Is this a clock? *Show students an analog clock.* **yes**

2. Please touch each number on the clock and count. *Have students start with their finger on 12.* **1, 2, 3, 4, 5, 6, 7, 8, 9, 10, 11, 12**

3. Move the hour hand to 6. *Make sure the minute hand stays on 12.* **Student should move the short hand to 6.**

4. Is this a half hour? *Move the minute hand from the 12 all the way around the clock to the 12 again.* **no**

5. Show me 8 o'clock. **Student should move the hour hand to the 8 and the minute hand to the 12.**

6. Is this a dime? *Show students a dime.* **no**

7. Is this a $1 bill? *Show students a $1 bill.* **yes**

8. Where is a $1 bill? *Put a pile of pennies, nickels, dimes, $1 bills, $5 bills, and $10 billes in front of student.* **Student should point to a $1 bill.**

9. Take away five cents. **Student should remove five pennies or one nickel.**

10. What is five pennies plus seven pennies? **Students should show or say any combination of pennies, nickels, and dimes that equal 12 cents.**

- **Minimal Understanding:** 0–3 of Questions 1–10 correct
- **Basic Understanding:** 4–7 of Questions 1–10 correct
- **Secure Understanding:** 8–10 of Questions 1–10 correct

Use the Student Assessment Record, page 142, to record the assessment results.

Weeks 31–32 Final Oral Assessment 2 (Intermediate and Advanced English Learners)

Directions: Read each question to the student, and record his or her oral responses. Some questions have teacher directions. Teacher directions are indicated in italics. Allow students to use pencil and paper to work their responses.

1. Point to a clock. *Show students several photos, one of which is a clock.* **Students should point to a clock.**

2. What time is it? *Show students an analog clock. Set it to 10 o'clock.* **10 o'clock**

3. Please count aloud as you move the hour hand forward three hours. **Student should move the hour hand forward from the 7 to the 10.**

4. Move the minute hand forward a half hour. *Give the student an analog clock with the hour hand on 8 and the minute hand on 12.* **Student should move the minute hand from the 12 to the 6 in a clockwise direction.**

5. Show me three hours after 4 o'clock. *Set the clock to 4 o'clock.* **Student should set the clock to 7 o'clock.**

6. How many $10 bills are there? *Display a set of five pennies, three nickels, four $1 bills, one $5 bill, and two $10 bills.* **two**

7. Did I add two $1 bills or subtract two $1 bills? *Take two $1 bills from the pile.* **Subtracted two $1 bills.**

8. Which has more value? *Display a $1 bill and a $10 bill.* **the $10 bill**

9. How do you know? **One $10 bill equals ten $1 bills.**

10. Order the money from greatest to least value. *Give student a $10 bill, a dime, a nickel, a $1 bill, a $5 bill, and a penny.* **Student should order the money as follows: $10 bill, $5 bill, $1 bill, dime, nickel, penny.**

- **Minimal Understanding:** 0–3 of Questions 1–10 correct
- **Basic Understanding:** 4–7 of Questions 1–10 correct
- **Secure Understanding:** 8–10 of Questions 1–10 correct

Use the Student Assessment Record, page 142, to record the assessment results.

Vocabulary Card Introduction

Blackline masters of the words included on the **SRA Math Vocabulary Cards** are included for reproduction on the following pages. Use them as suggested, or create your own variations, to enhance vocabulary development.

- **Flash Cards** Make flash cards with words on one side of the card and definitions on the other side.

- **Resource Cards** Make resource cards with the definitions, or an illustration, on the same side of the card as the word.

- **Matching** Match the words with their definitions, written on additional cards.

- **Matching Pairs** Place cards in columns and rows facedown, turning two cards over at a time to match word pairs.

- **Organize** Use graphic organizers to group words into categories.

- **Bingo** Use words students have studied. Call out definitions. Have students cover the words with counters until they get four or five words in a row.

- **Picture Cards** Create an illustration of the word as a visual reference.

$1 bill

$5 bill

$10 bill

add

after

angle

array

bar graph

before

circle

clock

cone

cube

denominator

difference

digit

dime

divide

early

eight

equal

even number

face

factors

first

five

four

fraction

greater than

half hour

hour

hundreds

late

less than

line

line of symmetry

minute

multiply

nickel

nine

numerator

odd number

one

ones

ordinal numbers

penny

prime number

pyramid

rectangle

remainder

second

seven

six

square

subtract

sum

ten

tens

three

triangle

two

zero

Name _____

Date _____

Number Cards (1–10)

1	6
2	7
3	8
4	9
5	10

Name _____

Number Cards (11–20)

15	20
14	19
13	18
12	17
11	16

Name _____

Date _____

Dot Set Cards (1–10)

Name _____ Date _____

Plus One, Minus One, Plus Two, Minus Two

+1	**−1**
+2	**−2**

Planet Equations

Follow the directions to write or draw the equation shown at each planet.

Mercury _____

Venus _____

Earth _____

Mars _____

Jupiter _____

Saturn _____

Uranus _____

Neptune _____

Name _____ Date _____

Addition for 10

$\left(\text{Red}\right) + \left(\text{Blue}\right) = 10$

○○○○○○○○○○   _____ + _____ = 10

○○○○○○○○○○   _____ + _____ = 10

○○○○○○○○○○   _____ + _____ = 10

○○○○○○○○○○   _____ + _____ = 10

○○○○○○○○○○   _____ + _____ = 10

○○○○○○○○○○   _____ + _____ = 10

○○○○○○○○○○   _____ + _____ = 10

○○○○○○○○○○   _____ + _____ = 10

○○○○○○○○○○   _____ + _____ = 10

○○○○○○○○○○   _____ + _____ = 10

○○○○○○○○○○   _____ + _____ = 10

Name _____ Date _____

Grocery Store

milk 5¢

grapes 1¢

bananas 3¢

carrots 2¢

apples 1¢

cherries 4¢

flour 5¢

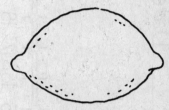

lemons 2¢

lettuce 4¢

mushrooms 2¢

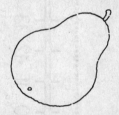

pears 3¢

tomatoes 3¢

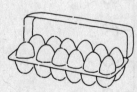

eggs 5¢

oranges 2¢

peas 1¢

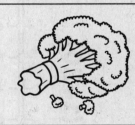

broccoli 3¢

corn 2¢

bread 3¢

Name

Date

Number Lines (1–20)

1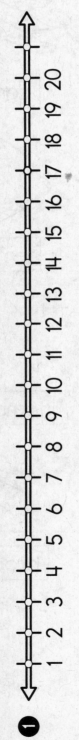
1 2 3 4 5 6 7 8 9 10 11 12 13 14 15 16 17 18 19 20

2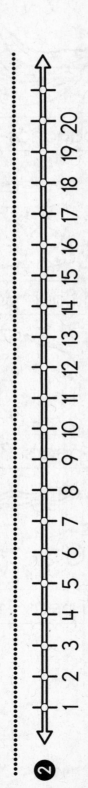
1 2 3 4 5 6 7 8 9 10 11 12 13 14 15 16 17 18 19 20

3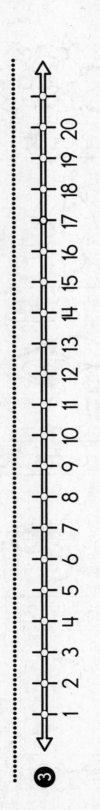
1 2 3 4 5 6 7 8 9 10 11 12 13 14 15 16 17 18 19 20

4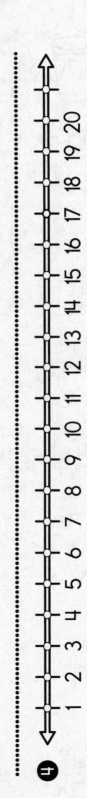
1 2 3 4 5 6 7 8 9 10 11 12 13 14 15 16 17 18 19 20

Name _____ Date _____

Plus, Minus, Equal

+	**−**	**=**
+	**−**	**=**

Comparison Table

Long	Longer	Longest
Short		
High		
Near		
Far		

Name _____ Date _____

Measuring Length

Write the object and its length on the correct lines.

A. Object: _____

   Length: _____

B. Object: _____

   Length: _____

C. Object: _____

   Length: _____

D. Object: _____

   Length: _____

E. Object: _____

   Length: _____

Name _____ Date _____

Map It
Follow directions to label these desks.
Answer questions about the map of desks.

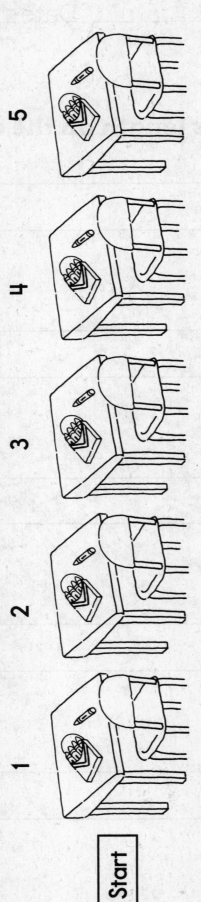

Start

Name _____ Date _____

Make-Your-Own Analog Clock

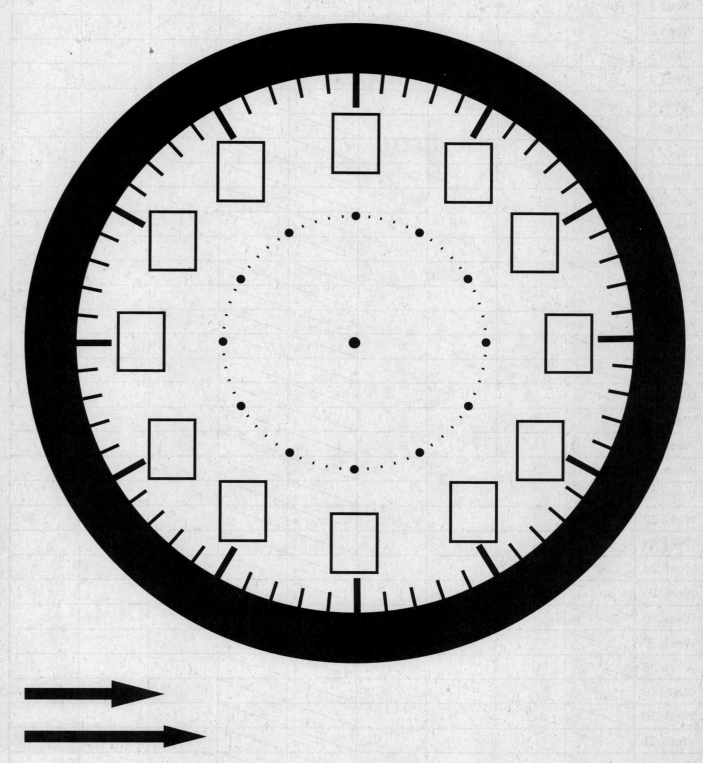

Copyright © McGraw-Hill Education. Permission is granted to reproduce for classroom use.

Student Assessment Record

	Individual Oral Assessment				Activity Completed	Final Assessment			
	Score	Beginning English Learner	Intermediate English Learner	Advanced English Learner		Score	Beginning English Learner	Intermediate English Learner	Advanced English Learner
Week 1	13				4				
Week 2	13				4				
Week 3	13				4				
Week 4	13				4				
Week 5	13				4				
Week 6	13				4				
Week 7	13				4				
Week 8	13				4				
Week 9	13				4				
Week 10	13				4				
Week 11	13				4				
Week 12	13				4				
Week 13	13				4				
Week 14	13				4				
Week 15	13				4				
Week 16	13				4				
Week 17	13				4				
Week 18	13				4				
Week 19	13				4				
Week 20	13				4				
Week 21	13				4				
Week 22	13				4				
Week 23	13				4				
Week 24	13				4				
Week 25	13				4				
Week 26	13				4				
Week 27	13				4				
Week 28	13				4				
Week 29	13				4				
Week 30	13				4				
Week 31	13				4				
Week 32	13				4